UNDERSTANDING TECHNOLOGY

by

R. Thomas Wright
Professor, Industry and Technology
Ball State University
Muncie, Indiana

and

Howard Bud Smith
Managing Editor—Technology
Goodheart-Willcox Company, Inc.

South Holland, Illinois
THE GOODHEART-WILLCOX COMPANY, INC.
Publishers

Library of Congress Catalog Card Number 88-31961
International Standard Book Number 0-87006-707-9

123456789-89-54321098

Library of Congress Cataloging in Publication Data

Wright, R. Thomas.
  Understanding technology.

  Includes index.
  1. Technology.   I. Smith, Howard Bud,
II. Title.
T47.W75 1989   600   88-31961
ISBN 0-87006-707-9

Introduction

What is technology? Why is it important in our lives? We need only look around to see how our lives are changed by technology.

We have tools and machines to make work easier. We have automobiles, hair dryers, and many other products to save us time and improve the quality of our lives.

Construction of all types provides shelter and convenience for all our activities. Houses and apartments keep us comfortable and safe from the elements. Bridges takes us over rivers. We can choose many different kinds of vehicles for travel.

Radios, telephones, and television keep us in touch with our friends, with our world. All of these advantages are the result of technology.

Technology is the knowledge of doing. It is a means of extending human abilities. Technology allows us to make useful products easier and better. Technology enables us to build structures on earth and, eventually, in space. Technology lets us move people and goods more easily.

UNDERSTANDING TECHNOLOGY introduces you to the various technologies. It explains the technologies as systems. These systems have inputs (such as people and materials), processes, outputs, goals, and constraints. You will be able to form opinions about the effects of technology. You will see that while most effects are good, others are not. As a result, you will be able to form opinions and make decisions about how to use technology wisely.

However, UNDERSTANDING TECHNOLOGY does more than *tell* you about technology. At the end of each chapter you will have a chance to apply what you have learned through carefully designed activities. In some you will build and test products. You may even use the product in competition with other students. In others you may be introduced first-hand to mass production or the use of tools. Another may ask you to investigate careers of your own choice and rate them against your own interests and expectations. We hope that this combination of information and "doing" activities will be a worthwhile experience for you.

R. Thomas Wright

Howard Bud Smith

UNDERSTANDING TECHNOLOGY explains the four basic industrial technology systems of Manufacturing, Construction, Communication, and Transportation. This book will explain technology . . . what it is, how it works, and how it affects all our lives.

Contents

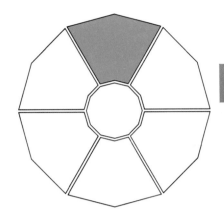

Becoming Familiar With Technology

This book is about the way we humans control our environment. If we are cold we can provide heat or heavier clothing. If we wish to be somewhere else we can provide faster and better methods of transportation. This ability to create change in our environment is the result of applying knowledge to solve problems.

As a result of these problem-solving activities, we are always changing the way we live, work, and move about. New inventions and discoveries are a part of our everyday lives.

What is new and startling today soon becomes commonplace. To our ancestors 40 thousand years ago fire was a new tool. It warmed them. They used it to fashion crude stone tools. While still vital to our lives, fire is now commonplace. Forty years ago, jet aircraft engines were just coming into use. They were exciting, fast, and powerful. Today, jet engines are common. They power most aircraft.

As you study Chapter 1, you will begin to see how important technology is to you. You will begin to see it as a series of carefully organized efforts. You will also begin to see that there is a right way and a wrong way to use technology.

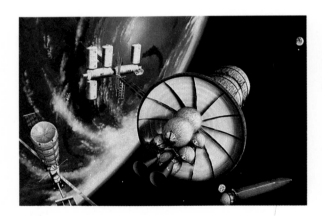

Can you imagine taking a ferry to the moon? This is an artist's idea of such a trip around the year 2000. The ferry (center) is powered by mixing liquid oxygen and liquid hydrogen. To the left of the space ferry is a space station as envisioned by some space planners.   (NASA)

Technology finds ways to capture the sun's heat. Pictured is a solar "furnace." Mirrors called "heliostats" concentrate the sun's rays. Then the rays are reflected onto the central tower where they heat water to create steam.

Chapter 1
What is Technology?

The information given in this chapter will help you:
- ☐ Define technology.
- ☐ Describe the difference between technology and science.
- ☐ Describe technology as a system.
- ☐ List and describe the parts of a system.
- ☐ Describe production and management processes.
- ☐ Discuss the goals of technology.

People have always wanted to live better. Early humans took shelter in caves and had hard lives. But they made progress by developing and using tools. They tied sharp rocks to sticks to make crude weapons. These tools let them hunt better, Fig. 1-1. Later they used tools to build houses from tree branches and logs.

Farming developed as the population of the earth grew. People could no longer depend on nature for food. There were not enough berries, roots, and wild animals to feed everyone.

The need to grow food led to the development of the plow, Fig. 1-2. This tool is often called the base for modern civilization. Without the plow we could not have efficient agriculture. Without the ability to grow large amounts of food we could not have cities. Without cities an industrial society cannot develop. An industrial society is needed to pro-

Fig. 1-1. Early humans used crude weapons to hunt game.

Fig. 1-2. The invention of the plow started efficient farming.   (Deere and Co.)

vide people with a wide range of products and services. This is the type of world in which we now live.

The development and use of tools continues. It causes our way of life to constantly change. This change is largely due to the tools we have developed and use. We no longer make crude stone and stick spears. Nor are plows a crudely shaped branch with a point. Our tools are now complex, Fig. 1-3. They are carefully designed and produced.

Still, they are made to serve people. They are used to make our lives easier. The main difference between our ancestors and us is in the tools we use.

WHAT IS TECHNOLOGY?

Tools are part of **technology** (teck-nol-o-gee). Have you ever used this word? Almost everyone has. But few people know what it means. *Technology is the technical means people use to improve their surroundings.* It is also a knowledge of using tools and machines to do tasks efficiently. We use technology to control the world in which we live. Technology is peo-

Fig. 1-3. Modern life depends on complex tools like this computer system. (Harris Corp.)

ple using knowledge, tools, and systems to make their lives easier and better, Fig. 1-4.

People use technology to improve their ability to do work. Through technology, people communicate better. Technology allows them to make more and better products. Our buildings are better through the use of technology. We travel in more comfort and speed as a result of technology. Yes, technology is everywhere and can make life better.

Almost everything in life has a good side and a bad side. Technology is no different. It can provide us with many goods and services. We can vacation in spots far from home because of efficient transportation systems. These same transportation systems bring us food, materials, and products from far away places. We can live in very cold climates because of advanced construction technology. We can use many exciting products because of effective manufacturing technology. By touching some buttons or dials we can talk to someone hundreds of miles away. Thanks to our communication technology, we watch or hear entertainment and news programs from around the world.

USING TECHNOLOGY WISELY

However, if technology is not used wisely it can cause damage. Poorly designed engines pollute the air we breathe. People driving off developed roads cause damage to the environment. Plants are killed and soil erodes more rapidly. Waste from factories can pollute the air, water, and soil. Fish die and the water is unsafe for human use. People can develop serious diseases from breathing polluted air. Food is contaminated by soil pollution.

Planes, cars, trains, and motorcycles can cause loud noise. They can pollute the silence many people want. A quiet, peaceful place becomes a noisy, irritating, location. *Amplified* (loud) music can cause permanent hearing loss.

Yes, improperly used, technology can cause damage. It can physically hurt people and the environment. New machines can cause jobs to become *obsolete* (out-of-date). Technology

change can force families to move to find new jobs. Friends and relatives may be left behind.

Also, if we don't understand it, technology can cause fear. Each of us is somewhat afraid of the unknown. Change is a threat. And technology causes change. Just think, many people who crossed the plains in covered wagons saw the first man to orbit the earth. Some people who rode horses to school are now in instant contact with the world through television and satellites. Later generations can remember when a 250 mile automobile trip was a major undertaking. They now can fly coast-to-coast in about five hours.

We like our life to be stable. We want to know tomorrow will be much like yesterday. But change is here and always has been. Only now, new technology make the changes come more quickly.

TECHNOLOGY AND SCIENCE

Often the terms, technology and **science,** are confused. Technology is said to be "applied

Fig. 1-4. Our lives are easier because people use technology to produce goods and services.

science." This is not true. Science deals with the natural world. It is the study of the natural laws which govern the universe.

Science tells us that objects will fall to the earth (law of gravity). Science explains why only certain plants are found on the Mojave Desert (plant ecology). Science tells us that steel exposed to oxygen will rust (chemistry). Science tells us that cross-pollinating plants will produce predictable results (biology). Science tells us that oil is most likely found near certain rock formation (geology).

On the other hand, technology deals with the human-made world. It is the study of ways people develop and use technical means — tools and machines. It tells us how to control the natural and human-made world. It is the study of the ways people use these technical means to transport, manufacture, construct, and communicate.

The difference between science and technology is easy to explain through an imaginary mountain hike. For this hike you arrive at the trail head in a manufactured vehicle. You traveled on a constructed road. You are wearing manufactured clothing. Technology was used to produce all these products.

However, during your hike you notice the moss growing on the north side of the trees. The trees are getting smaller as you climb. The glacier at the end of the hike is in a pocket on the east side of the mountain. In front of the glacier is a "green" lake. You may want explanations for the conditions you saw. Only science can provide the answer. They are the results of natural laws.

This is not to say science and technology are unrelated. Science deals with "understanding" while technology deals with "doing." Science helps us know why the world is the way it is. Technology helps us know how to do something efficiently.

SCIENCE FOLLOWS TECHNOLOGY

But both science and technology are very much connected. One causes the other to be explored. It is commonly thought that science generates technology. However, history tells us this is untrue. Technology comes first followed by science. Usually, people develop a technology (tool, machine, materials, etc.). Then the science behind it is explored.

For example, early humans melted metal to make tools. This moved the human race from the Stone Age into the Bronze Age. But these people knew nothing about melting temperatures for metals. They did not understand the properties of the metals they worked. They used technology without scientific understandings. Only much later did the science of metallurgy (metal properties) develop.

Likewise, the steam engine (a technological device) was in use 70 years before the theory of its operation (science) was explored. Machine tools (lathes, milling machines etc.) were developed before their mechanics (physics) was explained.

The Wright Brothers knew that wings had to be shaped in a certain way to get a plane to fly. However, the reason for these shapes was not known. They flew their airplane without a knowledge of *aerodynamics* (science of flight), Fig. 1-5. Similarly, the science of elec-

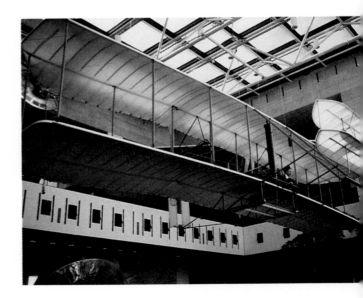

Fig. 1-5. The Wright brothers' airplane is an example of a technological advance.

tronics was developed after the first radio broadcast.

Technology, then, is the forerunner to science. People see a need. They develop a technical means of meeting that need. Later, other people explore the science behind the technological development.

TECHNOLOGY AS A SYSTEM

We often use the word system. We talk of our digestive system or the fuel system of a car. We read of computer and communication systems. Our homes are made comfortable with a heating system, and, in some cases, a cooling system.

All systems have some basic parts, as shown in Fig. 1-6. These parts are:
1. Inputs—the resources used by the system.
2. Processes—the actions taken to use the inputs.
3. Outputs—the result of the system.
4. Feedback—adjustments made to the processes to improve the outputs.
5. Goals—reason for the system.

Let us use a common system to describe these parts. Most of us have used a heating system, Fig. 1-7. The **goal** of the system is to make a home comfortable in cold weather. The system is designed to keep a house at an even, livable temperature.

The **input** to the system is a fuel source. The system may use natural gas, fuel oil, coal, or

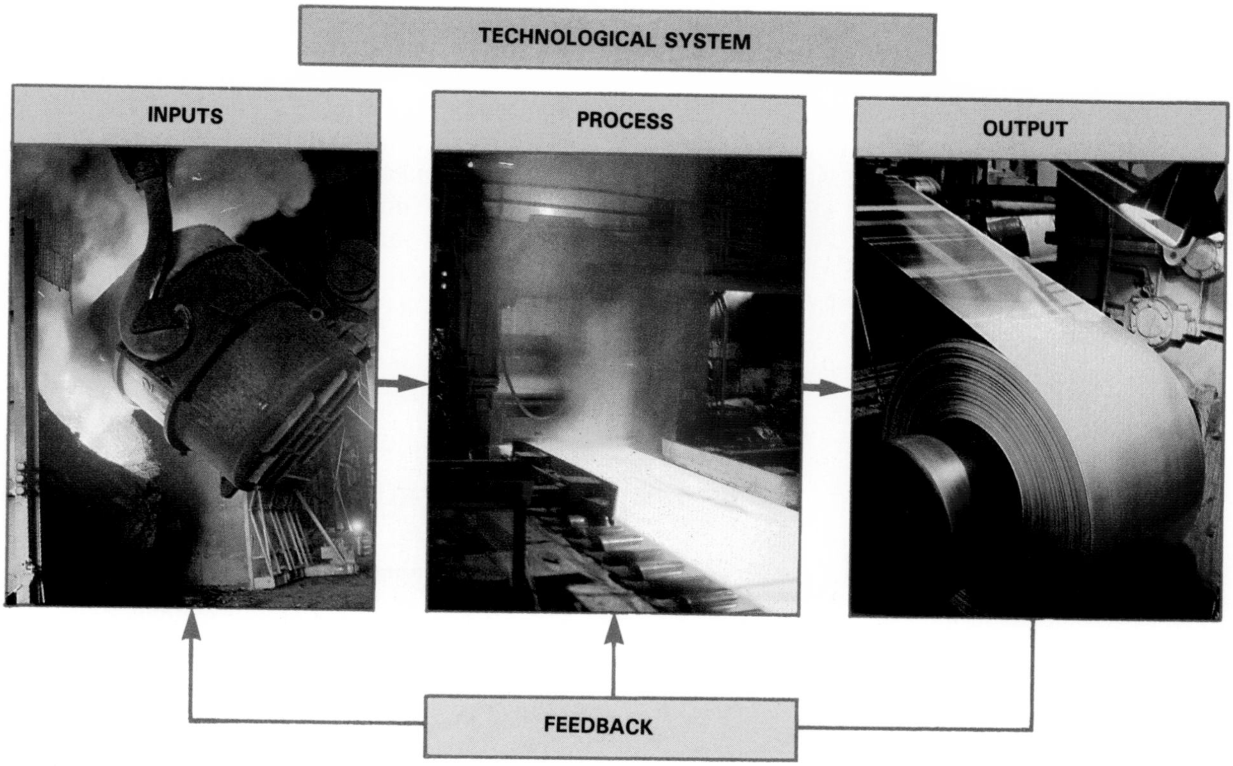

Fig. 1-6. Technological systems involve:
INPUTS. Left. Ingredients are being put into a steel furnace.
PROCESSES. Center. Steel is rolled into strips.
OUTPUTS. Right. Steel strip is being coiled.
FEEDBACK. Adjustments are made to improve the steel-making process. (USX Corp.)

wood as a fuel. Each of these inputs contain energy which can be converted into heat. The **process** used by the heating system is combustion. The fuel is burned. During burning the chemical structure of the fuel changes. This reaction gives off heat energy.

The heat energy is the desired **output** of the system. Other outputs include carbon dioxide gas and ashes. These are not needed but cannot be avoided.

An important part of the system is a thermostat. This device senses temperature. When the room falls below a set temperature the thermostat turns on the heating system. When the heat warms the room to the right temperature, the thermostat turns off the system. The thermostat is the **feedback** for the system. It provides data (information) which adjusts the system.

All technological systems have these same five parts. They have inputs, processes, outputs, feedback, and goals.

INPUTS TO TECHNOLOGICAL SYSTEMS

There are six major inputs used by technological systems, Fig. 1-8. These are:
1. People.
2. Natural resources (materials).
3. Capital (tools and machines).
4. Finance (money).
5. Knowledge (information).
6. Energy.

Each of these six inputs are essential for technological systems. They are all used and must be present.

People and Technology

Technology is developed by people. It is used by people. It is designed to serve people. Also, people are a very important input to technological systems.

People bring many skills and abilities to the technological system, Fig. 1-9. They may bring an ability to work with their hands. They may provide the *manual labor* for the system. Other people may manage the operation of the system. They may use their ability to organize tasks and direct the work of other people. Still other people may bring *technical knowledge* to

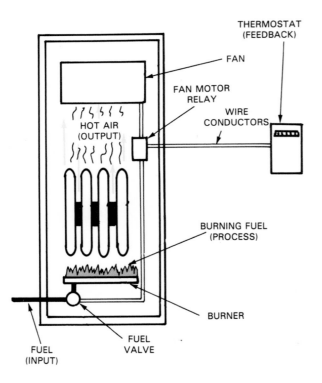

Fig. 1-7. A heating system is a technological system designed to keep buildings warm.

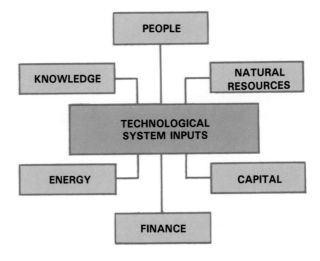

Fig. 1-8. The inputs to technological systems. Every system must have them.

the system. They may design products and machines. Also, they may organize the parts of the system into a whole. It takes many people with varying abilities to establish and operate technological systems.

Fig. 1-9. People bring their personal skills and knowledge to a task. These include the ability to (top) operate a machine, (center) design a product or system, and (bottom) manage a company. (Ohio Art Co.)

Materials and Technology

Technology involves employing technical means to use resources. A major resource used by all technological systems is materials, Fig. 1-10. These materials are gases, liquids, and solids.

Solid materials may be changed in shape and size by the system. They may be changed into products or become part of constructed works (buildings, roads, dams, etc.). They may be the fuels used to power the system.

Whatever their use, all material can be traced back to the earth. They are the natural resources found on or in the earth, in water, or in the air. Most of these materials (such as iron ore) are in limited supply. Once they are used they are gone. Some material are renewable. We can grow more of them. Trees, animals, and farm crops are examples of renewable materials.

Machines and Technology

All technological systems use capital goods. These are the human-made means of production. Capital goods are often called machines,

Fig. 1-10. Materials are a major input to technological systems.

Fig. 1-11. They are the machines used to cut and shape materials, broadcast messages, carry people and cargo, print publications, construct buildings and roads, etc.

Machines are present in all technological systems. They extend our ability to do something. Machines let us manufacture, construct, transport and communicate more efficiently.

Finances and Technology

We have all heard the saying, "It takes money to make money." The same is true for technological systems. It takes money to design,

engineer, and build technological systems. People must be paid to build and operate the systems. Machines must be built or bought. Materials must be purchased. Without money, no technological system could be built or operated.

Knowledge and Technology

Each technological system is based on knowledge. It is a knowledge of doing. But this knowledge varies with the technological activity. There is knowledge related to communication, construction, manufacturing, and

MACHINES AND
TECHNOLOGICAL SYSTEMS

COMMUNICATION

CONSTRUCTION

MANUFACTURING

TRANSPORTATION

Fig. 1-11. Communication, construction, manufacturing, and transportation technology systems use machines. (Harris Corp.; Caterpillar, Inc,; Ohio Art Co.,; CSX Corp.)

transportation. This knowledge might include information on material processing, audio (sound) and video (sight) recording, vehicle design, construction techniques, to name a few. The knowledge also varies with a technological system. For example, manufacturing technology requires many different types of knowledge. Some people will have the knowledge required to design products. Others will know how to develop manufacturing processes. Still other people will know how to operate manufacturing machines. Another group will possess the knowledge needed to service and repair manufactured goods.

Energy and Technology

All technological systems use energy. Each system uses machines which process inputs. Energy powers manufacturing machines, broadcasting equipment, transportation vehicles, construction equipment, and many other parts of technological systems. This energy comes from a number of sources including petroleum, natural gas, coal, wind, falling water, the sun, nuclear reactions, and wood.

PROCESSES OF TECHNOLOGY

Technological systems use two major types of processes. These, as shown in Fig. 1-12, are:
1. Production processes.
2. Management processes.

These two processes are united. They work together to convert the resources into the desired outputs.

Production Processes

Production processes are actions which change the inputs into outputs. These processes, as shown in Fig. 1-13:
1. Change materials into products (manufacturing).
2. Change materials and manufactured goods into constructed works (construction).
3. Change information into print or broadcast (media) messages (communication).
4. Change energy into power to move people and cargo (transportation).

The production processes are the technical means which are at the heart of each technological system. *They are the activities in which people use their knowledge to operate machines, powered by energy, to process materials or information or to transport people and cargo.*

Management Processes

Production processes do not run themselves. People must plan for their use. They must organize the workers, materials, and machines into an efficient system. They must direct and motivate people to operate the system properly. And they must evaluate the operation of the system. They must insure that the inputs are properly processed into outputs. These actions are called management. The people who complete these tasks are, likewise, called *managers.*

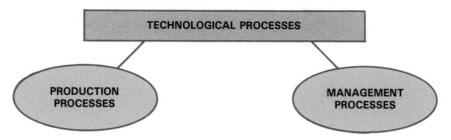

Fig. 1-12. The two types of technological processes are production processes and managerial processes.

Fig. 1-13. Production processes are used by four technological systems. A—To make products (manufacturing). B—To build structures (construction). C—To communicate information (communications). D—To transport people and goods (transportation). (Weyerhaeuser Co.; AMP, Inc.; CSX Corp.)

They plan and direct productive activities to insure they are efficient.

OUTPUTS OF TECHNOLOGICAL SYSTEMS

Almost all technological systems have two types of outputs, Fig. 1-14. The first is the one for which the system was designed. These outputs may be a manufactured product, constructed work, communicated message, or transported person. Each of these outputs are seen as good. They are what people want and need to live better.

However, most technological systems produce other outputs. These are in addition to the desired outputs. These other outputs may be of two kinds:
1. Scrap and waste.
2. Pollution.

Scrap and waste are materials left after the production processes are completed. Most of these materials do not harm the environment. They may be pieces of metal, wood, ceramics, or plastics, the exposed film not used for a

Fig. 1-14. The outputs of technological systems include (left) products and services we use and (right) scrap and waste. (Coachman Industries, American Petroleum Institute)

movie, or the empty shipping container after the product is removed.

Scrap is reusable. Waste must be discarded. All technological systems should be designed to keep scrap and waste to a minimum. However, they cannot be totally avoided. If you are cutting circles from sheets of plywood, waste will be present. Proper layout will limit the waste but cannot eliminate (stop) it.

Pollution is output which harms the *environment*. (The environment is our air, soil, and water — our surroundings.) Pollution can be chemicals, noise, unsightly structures, dust, smoke, etc.

Again, pollution is a part of many technological systems. The challenge is to control it.

Many of our activities directly create pollution. But we are trying to correct this. Automobiles put chemicals into the air. In recent years gasoline mixtures and automobile exhaust systems have been changed to reduce pollution. These actions have had some good effects on the environment. Steel mills have in-

stalled air filtering systems to reduce the pollutants they send into the air. Jet aircraft create noise. Now, new quieter engines reduce this noise. These are only a few examples of pollution control. They represent a start. We all have a great deal to do so that pollution is reduced. Households, as well as our industrial plants, must work hard at controlling pollution.

GOALS OF TECHNOLOGY

Most technological systems have two major goals. First there is the goal to meet human needs. This goal should be behind every technological activity.

A second goal exists when companies use the systems. This goal is **profit**. It is the reward earned by the owners for taking financial risks.

These goals are not opposed to one another. In fact, in the business world, both must exist. A company can only make profit when it meets human needs and wants.

What is Technology? 17

The best technological system will be useless unless it helps people live better. It must extend people's ability to control the environment.

SUMMING UP

Technology is the knowledge of efficient action. It is using technical means to extend human potential or ability. Every person is directly affected by technological systems. These systems provide the products we use, buildings we enter, communication media we read, hear, and view, and the transportation systems on which we travel. Technological systems use inputs which are used or processed into desirable outputs. The inputs are people, materials, machines, energy, knowledge, and finances. These enter into both production and managerial processes to produce the desired output. The system also produces undesirable outputs including scrap, waste, and pollution. Each of us should fully understand technology so we can control its use to improve our lives and protect the environment.

KEY WORDS

These words were used in this chapter. Do you know their meaning?

Feedback, Goals, Inputs, Outputs, Processes, Science, System, Technical means, Technological system, Technology.

ACTIVITIES

1. List all the ways you think a typical household pollutes our environment.
2. Divide a paper in half lengthwise. Write "science" on the right-hand column and "technology" on the left-hand column. On your way to school list those things you see that are explained by:
   a. Scientific law in the right-hand column; i.e. leaves changing color, tree dying in a vacant lot.
   b. Technology in the left-hand column; i.e. home being constructed, street being paved, etc.
3. Prepare a poster which will show the inputs, processes, outputs, goals, and feedback for a technological system.

TEST YOUR KNOWLEDGE
Chapter 1

Do not write in this text. Place answers to test questions on a separate sheet.

1. Select the statement which best defines technology:
   a. Development and use of tools.
   b. A way of making more and better products.
   c. Changing materials using tools.
   d. A knowledge of using tools and machines to do tasks efficiently.
2. Name four activities that technology helps us do better.
3. Technology is always good even when we do not use it wisely. True or false?
4. _____ is the study of natural laws which govern the universe.
5. Activity which develops technical means (tools and machines) to control the natural and human-made world is called _____.
6. Name and explain the separate parts of a technology system.
7. What do the following have in common: People, natural resources, capital, finance, knowledge, and energy?
8. Acts which change inputs into outputs are called _____.
9. _____ processes are activities which plan and direct productive activities to insure their efficiency.
10. List the two major goals of technological systems.

APPLYING YOUR KNOWLEDGE

Introduction

You have learned that technology is the use of tools, machines, or systems to make a job easier to do. The tool may be as simple as a shaped stick or as complex as a computer system. However each tool lets people do more work with less effort. In this activity you will first make a simple product with only your hands. Then you will use some special tools to do the same task. You will be able to see how tools extend your ability to do work.

Equipment and Supplies

Activity 1
Tote tray or pan
Rags or paper towels
25 lb. red clay (potter's clay)
Water

Activity 2
For each group of five students:
Materials from Activity 1
Rolling Sheets (approx. 20 — 1/8" x 12" x 12" tempered hardboard)
Rolling pin
Thickness strips (1/4" x 3/4" x 12" hardboard)
Cutting jig, see Fig. 1A.
   1 pc. 26 to 28 ga. x 2" x 16" galvanized steel
   1 pc. 26 to 28 ga. x 1" x 2" galvanized steel
Grooving jig, see Fig. 1B
   4 pc. 1/4" x 1/2" x 4 1/2" pine
   1 pc. 1/8" x 4" x 4" hardboard
Groove-forming tool (1/4" dowel with one end rounded)
Curing boards (approx. 50 — 1/4" x 5" x 5" hardboard)

Procedure:

Activity 1
1. Select a ball of clay approximately 3 in. in diameter.

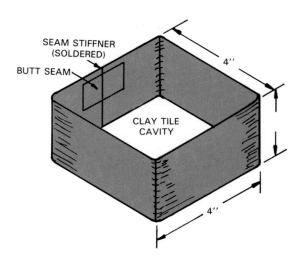

Fig. 1A. Tile cutter.

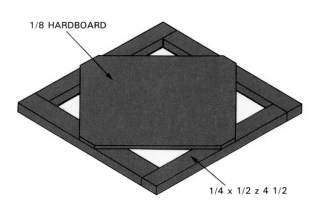

Fig. 1B. Template for forming the decorative grooves.

2. Form a decorative wall tile using only your hands. Look at the drawing, Fig. 1C, for the product. The tile is 1/4 in. thick and 4 in. square. It has a decorative groove across each corner. The groove is 1 in. in each direction from the corner.
3. Compare your tile with those made by other students.

Activity 2
1. Divide into groups of five students.

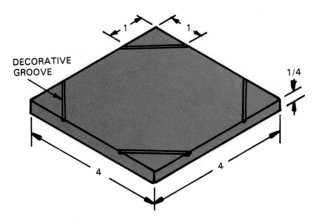

Fig. 1C. Decorative tile.

2. Assign a number to each member of the group from 1 to 5.
3. Each student will work with the tools or supplies listed:
   Student 1 — about 5 lb. of red clay and six rolling boards.
   Student 2 — rolling pin and two thickness strips
   Student 3 — tile cutter
   Student 4 — grooving jig
   Student 5 — 10 curing boards
4. Arrange yourselves around a workbench in numerical order.
5. Repeat the following tasks until you have made 10 tiles.
   Student 1:
   a. Wedge the clay into 2 1/2 to 3 in. balls. Wedging means pounding and rolling the ball on the bench top. This removes air bubbles.
   b. Place the ball on a rolling sheet and slide it to student 2.
   Student 2:
   a. Space the thickness guides about 5 in. apart. Place the clay ball between the strips.
   b. Roll the clay to proper thickness. Make sure that you produce a sheet which is about the same in width and length. Use a rolling pin over the thickness guides as shown in Fig. 1D.
   c. Remove the thickness strips. Slide formed clay and the rolling sheet to student 3.

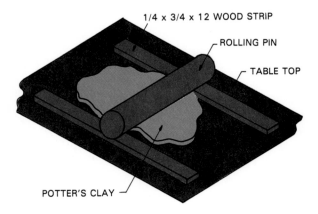

Fig. 1D. Rolling the clay.

Student 3:
a. Cut a 4 in. square from the clay sheet.
b. Remove the excess clay and give it to student 1.
c. Slide the clay square on the rolling sheet to student 4.

Student 4:
a. Place the grooving jig over the clay square. Use the dowel to form the grooves. You may need to dip the dowel in water before each forming pass.
b. Remove the grooving jig.
c. Slide the clay tile and the rolling sheet to student 5.

Student 5:
a. Carefully remove the tile from the rolling sheet.
b. Place the tile on a curing sheet.
c. Return the rolling sheet to student 1.
6. Write a short report which discusses how technology helped you make a product an easier way.

DEERE PLOW, 1838

AN ORIGINAL STEEL PLOW MADE BY JOHN DEERE
AT GRAND DETOUR, ILLINOIS

GIFT OF DEERE & COMPANY

We differ from our ancestors in the tools we use. The plow on the left required a great deal of manual labor. Now we control complex machines with knobs and buttons. (Deere and Co., Westinghouse Co.)

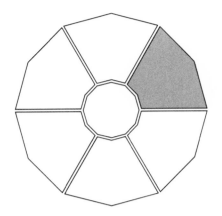

Resources and Technology

Resources are the many kinds of supplies, material, and services you draw upon to get anything done. Whether you are writing a letter, making a pizza, or building a house, you need resources.

Consider the simple process of writing a letter. What resources do you need?

First, there is the matter of tools or machines. Does letter writing require a tool? Of course! At the very least you will need a pencil. Then you may need a sharpener, another tool. But you may decide to use an electric typewriter, a machine.

You will need a chair and a surface for writing or typing. The chair and work table are made of materials too. Wood, metal, or plastic are common choices. The chair could be upholstered with cloth or fabric.

What other material will you need? Let us not forget the paper that will carry the message. Then you would use an envelope to protect the letter as it goes through the mails.

The mails! That means you'll need people from the post office to help you send the let-ter. What about making the paper? People cut the trees, carried them to the paper mill. Other people turned the wood into paper. If you are using imprinted stationery, someone did it on a printing press.

Then there is the need for energy. If you use an electric typewriter, energy powers the small motor that operates it. If you chose to write the same letter to 20 of your friends you might use a different machine. A personal computer would do the job faster. Again, electric power would energize the computer circuits.

Anything else? What about information? Surely, you had to have something to tell in the letter. Chances are you will tell something you heard, read, or experienced. So, information is a valuable resource.

Technology is the system that puts the resources to use. No matter what we are doing we use these same resources. You will see how various resources are tapped as you read the next five chapters. You will also come to realize that technology ties these resources to one another.

Computers rapidly process information for communication.

Raw materials are needed to produce lumber for furniture and pulp for paper. The resource comes from trees. What resource is the lumberjack providing? (Weyerhaeuser Co.)

Machines and people are needed to produce paper and imprint stationary. (GE)

Energy is required to process wood into paper. It is also needed to produce the electric power needed by an electric typewriter. Energy can be collected by pumping it from oil wells. It can also be supplied by solar collectors. (USX Corp.)

Chapter 2
Tools and Technology

The information given in this chapter will help you:
- ☐ Discuss humans as tool makers and tool users.
- ☐ Describe the difference between a tool, mechanism, and a machine.
- ☐ Describe the types of tools used in many areas of society.
- ☐ List, describe, and give examples of the six major types of primary tools.
- ☐ Describe and give examples of the six mechanisms or simple machines.
- ☐ Describe the use of the lever and the wheel and axle as force multipliers and distance multipliers.
- ☐ List and describe the major parts of a machine.
- ☐ Describe the major types of machines.

Humans are different from all other species of living things. People can reason: "If X is true, then Y must also be true." Or, "If I do this I can expect that to happen."

People also think of the future. They can plan to do things at some later date. They can plan their actions over a period of time.

Humans can also think of different ways to do the same task. People can adjust their behavior to meet different situations. Their minds "see" things and relationships.

All of these abilities are called rational thought. They can then use language to tell other people their thoughts.

Finally, humans have a sense of right and wrong. They can think, "I should do this but not that." This ability is called making moral judgments. All of these abilities are positive differences between humans and other mammals.

But all people share some drawbacks. People enter life without a "blueprint" of action. Cats know how to hunt rodents. Birds know how to build nests. Cows know how to care for their new-born calves. But humans have few of these built-in knowledges called instincts. People have to learn most of their actions.

Humans are not very well equipped physically for the world. Horses have hooves and teeth well suited for the grassy plains. Beavers have long teeth for felling trees and stripping bark. Ducks have oil-covered feathers to shed water. Bears have heavy fur to turn away the cold. But people cannot survive with only their natural "equipment."

TOOL USERS

This shortcoming has been overcome with a special human ability. This is the ability to design, make, and use tools, Fig. 2-1. **Tools** are devices which are used to do a specific task.

Fig. 2-1. Humans design, build, and use many tools and machines. (Caterpillar Inc.)

Over the years humans have made a wide variety of tools.

The "tool box" of early humans contained a few crude implements and weapons. There was a pointed stick to spear game and fish. A carefully selected tree limb served as a club. Later, properly shaped branches were used to plow the soil.

Today there are tools for every job. Have you ever seen skilled carpenters at work? They have tools for every task. There is a heavy hammer to drive stakes into the ground. Another one nails 2 x 4s together. Still another nails shingles in place. Each trade, craft, or job has its own set of tools. These tools can be understood better by looking at three major features shown in Fig. 2-2:

1. The tool, a simple device used to complete a job.
2. A mechanism, a basic device to adjust or power a tool.
3. A machine, a combination of tools and devices which can be used to complete complex tasks.

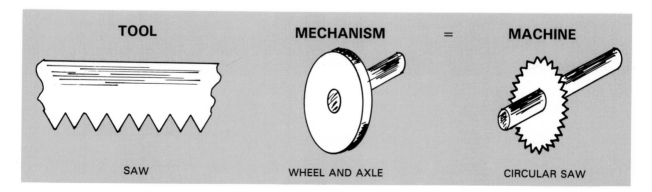

Fig. 2-2. People use tools, mechanisms, and machines to do work. (When we combine a tool and a mechanism, we have a machine.)

TOOLS

From earliest history, humans have survived by the use of their hands and arms. These body parts are controlled by an advanced brain. The hand held a rock to form a crude hammer. Later a stick handle was attached to the rock. It made a more efficient hammer. Using a sharp rock changed the hammer into a hatchet. In all cases the human-made device was a tool. And, as Henry Ward Beecher once wrote, a tool is but the extension of the human hand. See Fig. 2-3.

As we view tools we can see that there are many different types. There are:
1. Tools used with language: printing presses, type, for example.
2. Tools used in religion: vessels, special clothing, etc.
3. Tools used with commerce: scales, weights, money, etc.
4. Tools used by government: military weapons, police equipment, etc.
5. Tools used in art: paint brushes, sculpture chisels, etc.
6. Tools used in games and sports: balls, bats, etc.
7. Tools used in pure sciences: microscopes, chemical apparatus, etc.
8. Tools used in technological systems: communication, construction, manufacturing, and transportation equipment.
9. Tools used in managing companies: computers, word processors, etc.

All of these tools can be traced back to another type of tool, Fig. 2-4. We call them tools to make tools. Another name for them is primary tools. Without them, humans could not develop other tools and machines to make our lives easier and better.

Each of these groups of tools made the human hand and arm more efficient. Each extended the human potential to do a job.

Measuring Tools

Early humans lived in small groups and needed few tools. But, as the population grew, people formed larger groups. With this growth

Fig. 2-3. A tool is an extension of the human arm. This means that it makes the arm and hand able to do more or better work.

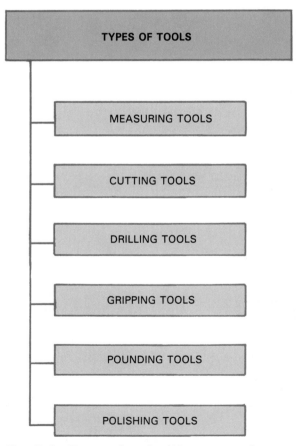

Fig. 2-4. Types of tools. We can use them to build other tools and mechanisms.

of civilization, people needed to measure things. They had to plot out fields to farm. They needed to measure trees to make houses. Grain placed in the village store houses had to be weighed.

Measurement became an important part of life. Humans use several different types of measurements. Volume and weight are important measurements for commerce. Gasoline is sold by the gallon (volume). Supermarkets sell hamburger by the pound (weight).

As we view the tools to make tools, two other types of measurement become important. People measure distances and relationships. We measure the size of objects (distance) and how one surface relates to another (squareness, etc.).

Measuring distances

Finding distance is a very important type of measurement. It tells us how far it is from one point to another.

A number of different tools measure distances. Some are considered precision measurement devices. They will give us very accurate measurements. Most precision measurement devices measure distances in thousandths (1/1000) of an inch or smaller. Other measure-

ment devices are nonprecision. They are used for normal or standard measuring activities. They give us measurements in *fractions* (1/8, 1/16, 1/32, etc.) of an inch, Fig. 2-5.

An automobile *odometer* will give us a standard measurement of long distances. Satellites using cameras and computers can measure the same distance within a few inches. This would be considered a precision measurement.

To measure shorter distances people often use a rule. This tool is a strip of material with measurement marks along its length. These marks will be in one of two systems:

1. U.S. Conventional—a measurement system in which the inch is the standard length. Twelve inches equals a foot and 36 inches equals a yard. Longer measurements are made in miles. A mile is equal to 5280 feet or 1760 yards.

2. SI Metric—a measurement system which uses the meter as its basic unit. The meter is divided into decimal parts. Common measuring units are the *centimeter* (1/100 meter), the *millimeter* (1/1000) meter and the *kilometer* (1000 meters). The metric system of measurement is now used by most countries in the world.

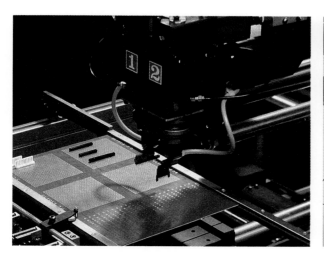

Fig. 2-5. We have tools that measure in thousandths of inches or finer. Others need not be so accurate. Left. These electronic circuits were designed with precision measurement. Right. The package designer here is using standard measurements. (AMP, Inc. and Ohio Art Co.)

The rule is often used to measure distances along a flat surface. It gives us length, width, and thickness measurements. Almost everyone has used a rule to make a measurement.

Sometimes we want to measure diameters. A common tool for this task is a caliper. The caliper is used in one of two ways. For inside diameters the caliper is fitted to touch the inside surfaces of a hole. For outside surfaces the tool is fitted to touch the outside surfaces of a shaft or rod. The distance between the caliper points is then measured by a rule. This tells us the diameter, Fig. 2-6.

The rule gives us standard measurements. *Precision* (very fine) measurements are often made with a micrometer. To use a micrometer the *workpiece* (part to be measured) is placed between the spindle and the anvil. The spindle is brought into contact with the part. The measurement is read on the micrometer barrel. Fig. 2-7 pictures some common measuring tools.

Measuring relationships

The relationship between two surfaces is often important. People want to know if a part

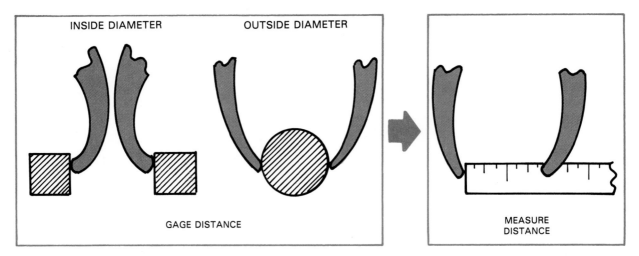

Fig. 2-6. Measuring with a caliper. Left. An inside caliper measures holes sizes. Center. An outside caliper checks outside dimensions of circular objects. Right. Rule being used to measure distance.

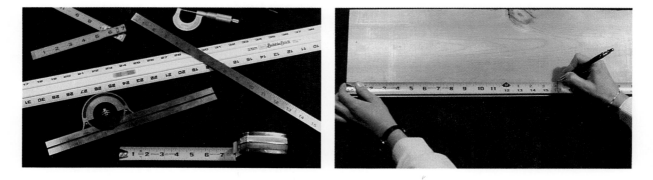

Fig. 2-7. Left. There are many kinds of common measuring tools. Right. A steel tape is used to measure a board to length.

is square. They need to know that a side is at *90°* (right angle) to an end. In another case it is very important that the ends of parts of a picture frame are at 45°. These measurements are made with tools called *squares,* Fig. 2-8.

A square is an "L" shaped tool. The two parts are exactly 90° to each other. The common ones are rafter, try, and combination squares. The combination square can measure both 90° and 45° angles.

Cutting Tools

Cutting tools remove material to size and shape parts. Each tool cuts away unwanted material until the part is shaped.

Cutting tools include three major types. These, as shown in Fig. 2-9, are:
1. Sawing tools.
2. Slicing tools.
3. Shearing tools.

Fig. 2-8. Squares are tools that measure angles. Note how a try square is used to mark a 90 degree angle on a board. Squares are often combined with rules.

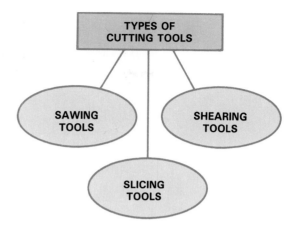

Fig. 2-9. These are the different families of cutting tools. Can you name examples of each type?

Sawing tools

A saw uses a tooth to cut the material. The tooth is a sharp, shaped projection (point) on a body. Typically the teeth are arranged along a strip or on a disc.

The saw must move to cause a cutting action. All handsaws are strips of metal with teeth along their edges. The saw reciprocates (moves back and forth) over the material to make the cut. Handsaws usually cut on the forward stroke. Typical handsaws, Fig. 2-10, are the woodworking crosscut, rip, and backsaws. Those designed for metal cutting are called hacksaws.

Disc (circular) saws are used with cutting machines. These and other machines will be discussed later.

Slicing tools

Slicing tools use a sharp, wedge-shaped edge to separate the material. The wedge cuts away unwanted material in the form of shavings. Typical slicing tools are the knife, chisel, and woodworking plane, Fig. 2-11.

Shearing tools

Shearing tools fracture (break) material between two opposing edges. The workpiece is placed between the edges (knives). The knives coming together cause the material to separate. Common shearing tools, Fig. 2-12, are tin snips and scissors.

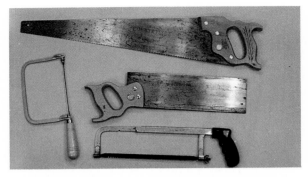

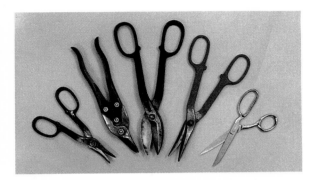

Fig. 2-10. Top. Different kinds of handsaws. Bottom. A crosscut is a type of handsaw used to cut lumber across the grain.

Fig. 2-12. Top. Shearing tools use two sharp opposing edges to fracture (cut) the materials. Bottom. A tin snips cuts sheet metal.

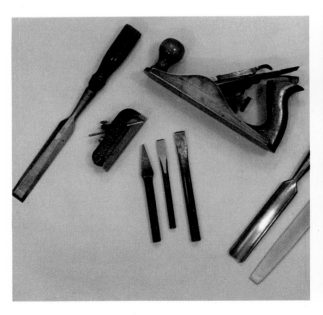

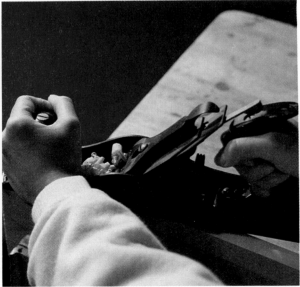

Fig. 2-11. Typical cutting tools include planes, chisels, and files. Hand planes are used to smooth board surfaces.

Drilling Tools

One of the first tools ancients developed was the drill. The first one was no more than a sharp-pointed stone. It was attached to a wooden shaft. The user rotated the shaft by holding it in the palm of the hands. The stone tip, pointed downward, rested on the piece to be drilled. When the hands were rubbed back and forth the shaft would rotate. First it would turn one way, then the other. At the same time, downward pressure forced the stone point into the work.

Today, drilling still uses the same action. A steel shaft has a cutter on its end. It rotates. Downward pressure forces it into the workpiece. This produces a hole.

Common drilling tools are the twist drill, the spade bit, and the auger bit. These are often held and rotated in either a hand drill or a brace, Fig. 2-13.

Gripping Tools

The hand is a natural gripping device. People hold and twist things with their hands. But the hand can only exert a limited amount of force. Therefore, tools for both holding and gripping have been developed.

Holding tools

Holding tools hold an object in place. They squeeze the part and keep it from moving. Typical holding tools are pliers, vises, and clamps, Fig. 2-14. These tools adjust to a wide range of sizes.

Turning tools

People often need to turn objects to position them or tighten a fastener. Nuts are turned onto bolts. Screws are turned into wood. Shafts are turned to align them properly.

Turning tools perform these tasks. We use wrenches, screwdrivers, and pliers (grip and turn). Wrenches are made in fixed sizes or are adjustable. Fixed-size wrenches usually come in sets to fit common bolts and nuts. They are either open-end or box styles. Adjustable open-end wrenches and pipe wrenches, Fig. 2-15, are most common.

Screwdrivers are available in several lengths and blade sizes. They also are made to fit different head shapes. Fig. 2-16 shows standard slot and Phillips head screwdrivers.

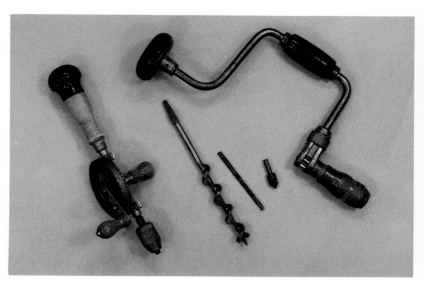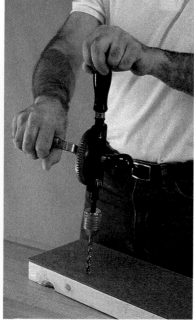

Fig. 2-13. Left. A sample of drilling tools. Right. Drills must be rotated to make them cut.

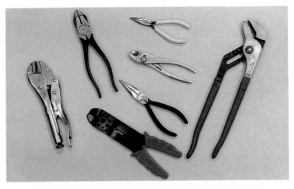

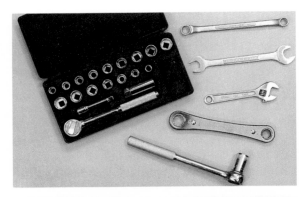

Fig. 2-14. Holding tools multiple the squeezing force of the hand. There are many types. Perhaps you have used some of these.

Fig. 2-15. These tools are useful for assembling parts using bolts and nuts.

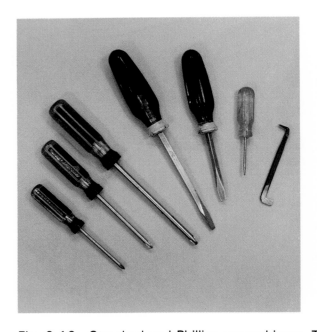

Fig. 2-16. Standard and Phillips screwdrivers. They turn screw fasteners into parts.

Pounding tools

From early times, people needed to apply force to objects. They needed to pound them into the ground or shape them. This need gave rise to pounding tools. Typically, they consist of a heavy head attached to a handle.

Today, we call these tools hammers. They come in many types and styles. To name a few, we have:
1. Claw hammers for driving all types of nails.
2. Riveting hammers for setting rivets.
3. Sledge hammers for driving stakes.
4. Shingling hammers for attaching roof shingles.
5. Tinners' hammers for closing sheet metal seams.
6. Mallets for driving a woodworking chisel.
7. Ball peen hammers for striking a cold chisel.
8. Rubber or plastic mallets for striking parts to align them.
9. Cross peen hammers for forging.

Fig. 2-17 shows some hammer types. How many do you recognize?

Polishing tools

Many items require a smooth attractive surface. Their parts are smoothed and polished using scrapers and abrasive grits. These tools remove small amounts of materials to improve the surface of the material. Hand and cabinet scrapers use a *curled edge* (burr) to scrape away the unwanted material.

Sanding and grinding tools use mineral grit to put uniform scratches into the material. The action replaces dents and large scratches with small, straight scratches. As these scratches become very small, the eye does not notice them. The eye and the hand tell us its smooth. Actually there is still a degree of roughness. But the surface is much smoother than it was.

Typical abrasive tools are loose and sheet abrasives, sharpening stones, and buffing compounds.

MECHANISMS

Tools served early humans well. But civilization grew. Demands for more food and products also grew. People using hand tools could not meet the demand. Technological advancement was necessary. Machines to produce more and better goods were needed.

But there is no direct step from tool to machine. The tool must be combined with a mechanism to produce a machine. What is a **mechanism?** *It is a basic device that will control or add power to a tool.* Science calls these

Fig. 2-17. Some of the many hammers we use. They are used for pounding and driving certain types of fasteners.

mechanisms simple machines. These simple machines or basic mechanisms multiply the force applied or distance traveled. An understanding of them helps us understand more complex machines.

There are six simple machines which work off of two basic principles, Fig. 2-18.

PRINCIPLE	SIMPLE MACHINE
Lever	Level Wheel and Axle Pulley
Inclined Plane	Inclined Plane Wedge Screw

Fig. 2-18. The six simple machines are based upon two basic principles.

Lever

Have you ever seen someone move a heavy box using a crowbar? If you have, you have seen a **lever** in action, Fig. 2-19. A lever, like all simple machines, is a force or distance multiplier. It can increase the force applied to the work. It makes us stronger than we really are. This is using a lever as a **force multiplier.**

A lever can also let us change the amount of movement created. A small amount of movement at one end will produce large movement at the other end. This uses a lever as a distance multiplier.

A lever is a device consisting of a lever arm and a fulcrum. The fulcrum is a pivot point upon which the lever arm rotates.

There are three basic arrangements for:
1. The lever arm.
2. The fulcrum.
3. The load to be moved.
4. The force to be applied.

These arrangements, as seen in Fig. 2-20, are called classes of levers. The principle of the first-class lever is used by scissors and pry-bars. Wheelbarrows and a hand truck are devices which use second class levers. When you use a broom or a baseball bat you are using third class levers.

Fig. 2-21 shows using levers as force and distance multipliers. In both cases a first-class lever is shown. The effects of load applied to these levers is as follows:

 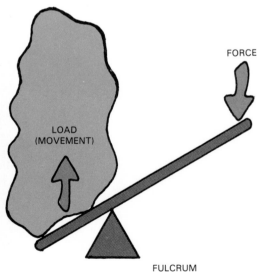

Fig. 2-19. Levers multiply force. The bar at left is an application of the lever illustrated at right.

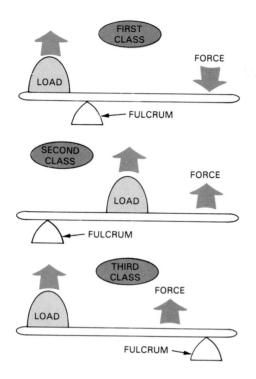

Fig. 2-20. The three classes of levers. Try to think of tools you use that fit these classes.

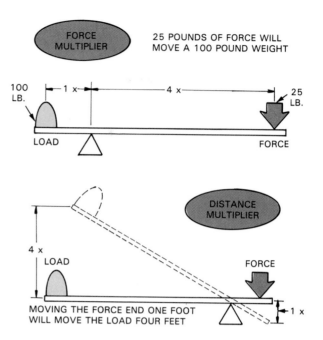

Fig. 2-21. Using a lever to multiply force and distance. Top. A 25 lb. force is multiplied four times. Bottom. Great force applied at right moves a short distance to move load four times farther.

FORCE MULTIPLIER (Fig. 2-21, top): The fulcrum is located near the load to be moved. Thus, a small amount of force, moving a greater distance, will move a load a shorter distance than the force movement.

DISTANCE MULTIPLER (Fig. 2-21, bottom): The fulcrum is located near the force applied to the lever. Thus, a force moving a short distance can cause a load to be moved a much greater distance. Of course, a much larger force must be applied to multiply the movement of the load.

Look around the school laboratory for examples of levers in use. Did you identify pliers, tin snips, claw hammers (when pulling a nail), etc.?

Wheel and Axle

A **wheel and axle** is a shaft which is attached to the center of a disc, Fig. 2-22. This mechanism operates a second class lever. The axle is the fulcrum and the load is applied to the wheel or axle.

If the force is applied to the axle, the mechanism becomes a distance multiplier. One

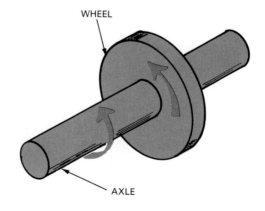

Fig. 2-22. The wheel and axle are an application of the lever.

revolution of the axle will cause the wheel to rotate one time. However the *circumference* of the wheel (distance around) is greater than the axle. Therefore the mechanism will move a greater distance. This action is used by bicycle drives and automotive *differentials* (gears that turn the axle).

If the force is applied to the wheel, the mechanism is a force multiplier. A screwdriver is a good example of this action. Try to drive a screw by gripping the shaft (axle) of a screwdriver. Then repeat the task gripping its handle (wheel). You will find that it is much easier to turn the mechanism by applying the force to the wheel. This principle is used for automobile steering wheels, other control knobs and wheels, and woodworking braces.

Study Fig. 2-23 carefully. Note how wheels and axles can be used as both force and distance multipliers. Can you think of other examples?

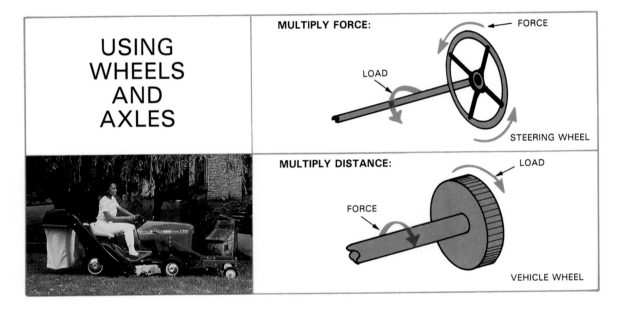

Fig. 2-23. Like the lever, the wheel and axle can be either a force or a distance multiplier.

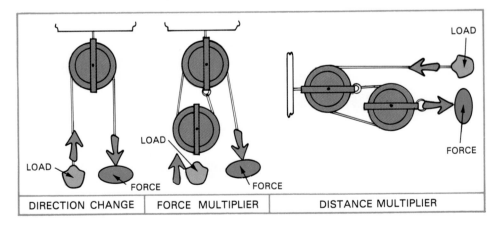

Fig. 2-24. Pulleys change direction, multiply force, and multiply distance.

Pulley

A **pulley** is a wheel with a grooved rim. It is attached to a loose axle. Pulleys are used by themselves or in sets. They can do three things:
1. Change the direction of force.
2. Multiply force.
3. Multiply distance.

Look at Fig. 2-24. See how pulleys are used to do each of these jobs.

Inclined Plane

An **inclined plane** is a mechanism using a sloped surface, Fig. 2-25. It operates on the principle that moving up a slope is easier than lifting straight up.

A simple experiment will test this principle. Pull a smooth weight up a slope. Use a scale to measure the force. Then lift it. You will find that the longer the slope the easier it is to move the object.

Wedge

A **wedge** is a set of two inclined planes, Fig. 2-26. This mechanism is used in many simple hand tools. The wood chisel, knife, ax, splitting wedge (for fire wood), and cold chisel operate on the wedge principle. Also, nails are wedges.

Screw

A **screw** is actually an inclined plane wrapped on a round shaft. The threads move slowly up the shaft as they go around it. The screw is a great force multiplier. It takes a great deal of rotating motion to move a nut a short distance onto a bolt.

Consider a 1/2 in. by 12 bolt (1/2 in. diameter with 12 threads per in.). To move the nut an inch the bolt must be turned 12 times. A point on the circumference of the bolt would move almost 19 in.

Fig. 2-26. Using a wedge to split a log. Do you see that it is two inclined planes put together?

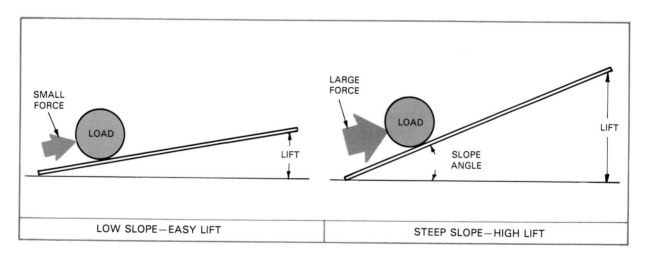

| LOW SLOPE—EASY LIFT | STEEP SLOPE—HIGH LIFT |

Fig. 2-25. The inclined plane. Load can be lifted with less force. The lower the slope, the easier the load is to move.

MACHINES

Each type of technology uses **machines.** There are machines used in communications. Printing presses produce newspapers and magazines. Transmitters send signals through the airways to radios and television sets. Switching gear interconnects our telephones.

Machines are used in transportation too. Trucks, trains, ships, and airplanes move people and cargo. Conveyors are used to load and unload cargo from vehicles. Computers maintain reservation information.

Machines are used in construction. Bulldozers prepare construction sites. Cranes lift structural steel into place. Cement mixers prepare concrete for use. Power trowels smooth the concrete while saws cut joints into it.

Machines are used in manufacturing. Presses stamp out parts. Welders are used to fuse metal. Robots move parts from place to place.

Many of these machines will be presented in Section 3 of this book. But all of these machines can be traced back to other types of machines. They are a special type of manufacturing machine. They are the machines which make machines. They are called machine tools. Without them there would be no communication, construction, general manufacturing, or transportation machines. Machine tools are used to change raw materials into parts.

The parts later become machines and other products.

Machine tools have four major elements, as shown in Fig. 2-27. They have:
1. A basic structure (frame, bed, table, etc.).
2. A power unit (electric motor, hydraulic drive, etc.).
3. Control unit (feed, speed, and depth of cut controls).
4. Tool (device to produce a cut).

Machine tools can be grouped into six major classes. These, as shown in Fig. 2-28, are:
1. Turning machines.
2. Drilling machines.
3. Milling and sawing machines.
4. Shaping and planing machines.
5. Grinding and sanding machines.
6. Shearing machines.

Each machine tool has its own way of operating. It cuts materials into shapes using different motions and tools.

Turning Machines

Turning machines were one of the first machines to be developed. They are almost as old as civilization itself. The potter's wheel is an example of an early turning machine. Later turning machines are the lathes.

Turning machines use a stationary tool. The material to be shaped is rotated around an axis.

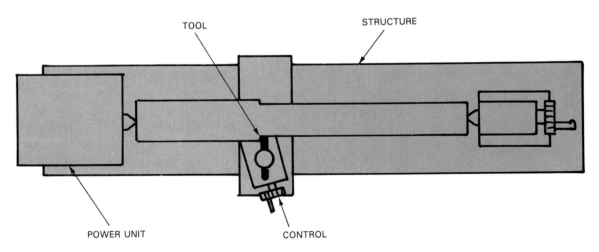

Fig. 2-27. Every machine tool has these four basic parts.

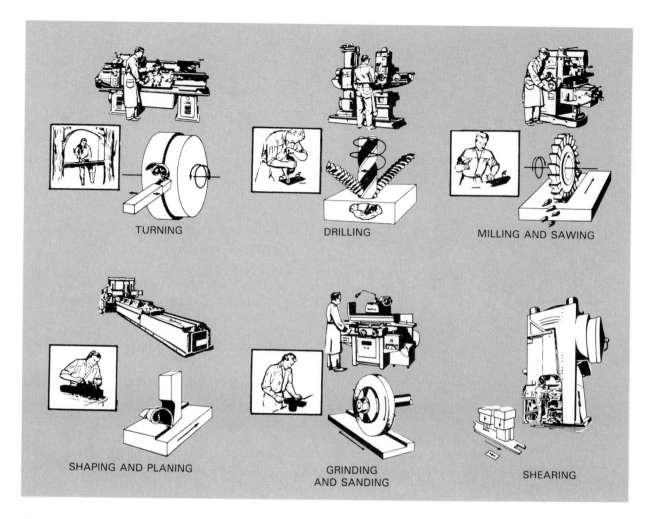

TURNING

DRILLING

MILLING AND SAWING

SHAPING AND PLANING

GRINDING AND SANDING

SHEARING

Fig. 2-28. All machine tools can be classified under one of these six basic types.

The stationary tool is fed into the work. The cut is created by slowly moving the tool along or into the rotating work.

Drilling Machines

The first drill was probably invented over 40,000 years ago. You learned earlier that it was no more than a pointed stone on a shaft.

Today many holes are produced by clamping a drill bit in a chuck (tool holder). The chuck is rotated and pushed downward. This causes the rotating drill to feed into the work and cut a hole. The most common drilling machine is the drill press, Fig. 2-29.

Milling and Sawing Machines

Milling and sawing machines use either the straight or circular saw blades discussed earlier. A motor makes the blade move. The work is fed into the moving blade, creating a cut.

Many of these machines use rotating circular blades or cutters. The most common machines of this type are the milling machine, table saw, and radial saw. The woodworking surfacer, jointer, shaper, and router use the same cutting action.

Power hacksaws, scroll (jig) saws, and saber saws use reciprocating straight blades. The band saw, Fig. 2-30, uses a straight blade which

Fig. 2-29. A drill press is the most common of drilling machines.

has been welded into a loop. The blade travels around two wheels. This produces a *linear* (straight line) cutting motion at the workpiece.

Planing and Shaping Machines

This type of machine is generally limited to cutting metals. Both machines use a single point. The shaper moves the tool into the work to produce the cut. The metal planer moves the work into the tool. Both machines produce a flat cut on the surface of the work.

Grinding and Sanding Machines

These machines use an abrasive to cut material from the workpiece. The abrasives can be bonded into wheels or on a backing for sheets, discs, and belts. Generally, the work is moved against the moving abrasive. Grinders and sheet, disc, and belt sanders are the common machines in this group, Fig. 2-31.

Shearing Machines

Shearing machines slice materials into parts. They use opposed edges to cut the workpiece. The material is placed between the cutting edges (knives or blades). One edge is moved down forcing the material against the second edge. As more force is applied, the material is frac-

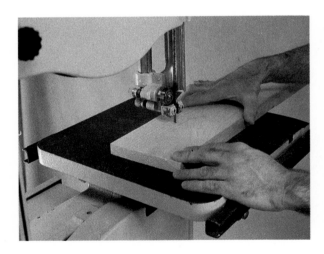

Fig. 2-30. A band saw uses a blade that rotates as one continuous loop.

Fig. 2-31. Sanding machines cut and smooth material.

tured (cut). A pair of scissors uses a shearing action. Common shearing machines are the sheetmetal shear, punch press, and paper cutter.

SUMMING UP

Technology is the use of technical means to extend human potential. A major component of technological action is the use of tools. Tools are devices humans have invented to do a job. Tools may be classified as measuring, cutting, drilling, gripping, pounding, and polishing tools.

But tools are used by hand. Hand work may be fun but it is slow. The demand for more and better products, communication media, buildings, and transportation devices gave rise to machines. Humans found that some basic mechanisms could be combined with tools to create these machines. They joined the lever, wheel and axle, pulley, inclined plane, wedge, and screw with the basic tools. From these came turning, drilling, milling and sawing, shaping and planing, grinding and sanding, and shearing machines. These machines can be used to make all other machines. From them come communication, construction, manufacturing, and transportation machines.

KEY WORDS

These words were used in this chapter. Do you know their meaning?

Distance multiplier, Force multiplier, Inclined plane, Lever, Machine, Mechanism, Pulley, Screw, Tools, Wedge, Wheel and axle.

ACTIVITIES

1. List six tools that you use at home and explain how they work.
2. List two machines that you have used or seen. Classify them in one of the six groups of machines.
3. Take a drawing of a machine your teacher gives you. Circle and name as many mechanisms as you can find that are included in the machine.
4. Build a simple item using tools. List each tool you use and the family of tools to which it belongs.

TEST YOUR KNOWLEDGE
Chapter 2

Do not write in this text. Place answers to test questions on a separate sheet.

1. Humans are different from other species of living things because (check correct answer):
   a. They walk upright.
   b. They can adjust their behavior to different situations.
   c. They can design, make, and use tools.
   d. Have to depend on tools to control their environment.
   e. All of the above.
   f. None of the above.
2. _____ are devices used to do a specific task.
3. Why is it said that a tool is an extension of the human hand?
4. List the six primary tools.
5. Which of the following tools would you use for measuring a diameter to the nearest thousandths of an inch:
   a. A ruler.
   b. A caliper.
   c. An odometer.
6. Indicate which of the following four measurements is the smallest metric measurement.
   a. Meter.
   b. Millimeter.
   c. Kilometer.
   d. Centimeter.
7. A _____ is a basic device that will control or add power to a tool.
8. The lever, wheel and axle, pulley, inclined plane, wedge, and screw are all examples of what?

9. A force multiplier is a lever which can _____ the force applied to the work.

10. A lever which produces large movement at one end when a force is applied to the other end is being used as a _____ _____.

11. If you turn the steering wheel on a car, you are using the wheel to:
   a. Multiply force.
   b. Multiply distance.
   c. Neither.
   d. Both a and b.

12. A pulley can be used to change the direction of force, multiply force, and multiply distance. True or false?

13. Which of the following are major elements of a machine tool?
   a. Frame.
   b. Tool.
   c. Electric motor.
   d. One of the controls.
   e. The part being made.
   f. The power source.

14. List the six major classes of machine tools.

Tools can be simple and small or complicated and huge. Left. This large woodworking machine is used in a cabinetmaking shop. Right. Pneumatic (powered by air) hammer drives nails on a construction site.   (Conestoga Wood Specialties Inc., and Paslode Co.)

APPLYING YOUR KNOWLEDGE

Introduction

You have read about tools and technology. Also, your teacher has told you about tools. Now you are ready to use this knowledge. Your teacher will show you how to put this knowledge to work. You will be shown how to use tools to extend your ability to do a job.

In this activity you will have common hand tools to help you build a simple game. You will be using common measuring, cutting, drilling, pounding, and polishing tools to make a tic-tac-toe board, Fig. 2A.

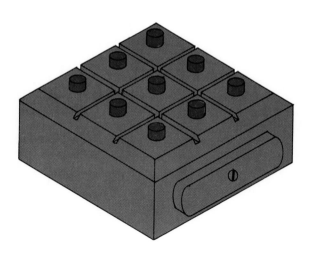

Fig. 2A. Tic-tac-toe.

Equipment and Materials

2 x 4 (1 1/2" x 3 1/2") construction lumber
1/4" x 3/4" wood strips
3/8" dowels
3/4 " x No. 6 flat head wood screws
Steel rule
Try or combination square
Crosscut or back saw
Miter box and handsaw
Hand drill
Brace
1/2" auger bit
1/16", 9/64", and 13/32" twist drills
Countersink bit
Flat wood file and/or rasp
Block or smooth plane
Abrasive paper and sanding blocks
Screwdriver
Scratch awl or center punch
Hammer or mallet

Product Drawing and Bill of Materials

Qty	Description	Size	Material
1	Game Board	1 1/2 x 3 1/2 x 3 1/2	Spruce or Hemlock
1	Peg Cover	1/4 x 3/4 x 2 1/2	Pine
8	Pegs	3/8 Dia. x 5/8	Birch Dowel
1	Screw	3/x × No. 6 Flat Head	Plated Steel

Fig. 2B. Bill of materials.

Procedure

The procedure for the tic-tac-toe is numbered from 1 to 43 but you do not have to complete the steps in that order. You may make the pegs, peg storage hole cover, and game board in any order. Each number is used only once so that you and your teacher can easily refer to a specific step without confusion.

Preparing to make the product

1. Study the drawings for the tic-tac-toe game, Fig. 2C.
2. Read the procedure for making the game.

3. Carefully watch your teacher as he/she demonstrates how to make the game.

Making the game board
Selecting and laying out the material:
4. Select a length of 2 x 4 construction lumber. (Note: the actual size is 1 1/2 in. thick by 3 1/2 in. wide.)
5. Lay out a line 3/8 in. from one end.
6. Lay out a line 3 1/2 in. from the first line.

Cutting out the game board:
7. Cut the end off to the outside of the 1/2 in. line to square the end of the board.
8. Cut off the game part barely leaving your line on the part.

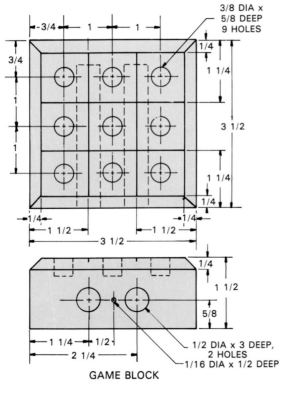

GAME BLOCK

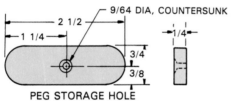

PEG STORAGE HOLE

Fig. 2C. Working drawings.

Laying out the game board. (See Fig. 2C and Fig. 2D.):
9. Draw lines 1 1/4 in. in from the edges and ends of the block.
10. Draw lines 1/4 in. in from the edges and ends.
11. Draw lines 1/4 in. down from the face on the edges and ends.
12. Locate and mark the nine peg holes.
13. Locate and mark the peg storage holes on one edge.
14. Locate and mark the peg cover pivot screw anchor hole.

Producing the game board:
15. Saw kerfs (shallow slots) about 1/8 in. deep on the four 1 1/2 in. lines.
16. Drill the nine 13/32 in. peg holes, 3/8 in. deep.
17. Drill the two 1/2 in. peg storage holes, 3 in. deep.
18. Drill the 1/16 in. pivot screw hole.
19. File and/or plane the 1/4 in. by 1/4 in. chamfers around the top of the block.
20. Sand all surfaces.

Making the pegs
Selecting and laying out the material:
21. Select a length of 3/8 in. dowel.
22. Check the end to see that it is square.

Producing the pegs:
23. Set a stop block on the miter saw for a 5/8 in. cut.
24. Cut the end of the dowel square, if necessary.
25. Cut eight pieces of dowel, 5/8 in. long.
26. Sand and lightly break (round) the ends of the pegs.

Making the peg storage hole cover
Selecting and laying out the material:
27. Select a length of 1/4 in. by 3/4 in. pine.
28. Draw a line 1/4 in. from the end.
29. Draw a line 2 1/2 in. from the first line.
30. Locate and mark the pivot screw hole.
31. Lay out the radius on each end.

Producing the peg storage hole cover:
32. Cut the end off to the outside of the 1/4 in. line to square the end of the board.

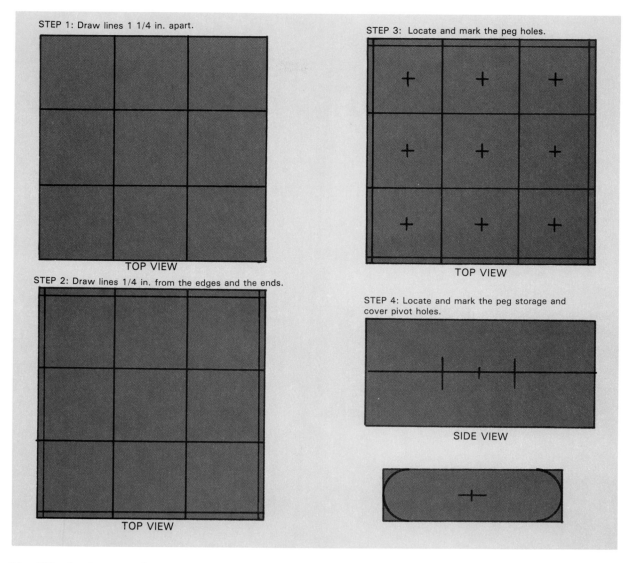

STEP 1: Draw lines 1 1/4 in. apart.

TOP VIEW

STEP 2: Draw lines 1/4 in. from the edges and the ends.

TOP VIEW

STEP 3: Locate and mark the peg holes.

TOP VIEW

STEP 4: Locate and mark the peg storage and cover pivot holes.

SIDE VIEW

Fig. 2D. Laying out the game board.

33. Cut off the hole cover barely leaving your line on the part.
34. Drill a 9/64 in. pivot screw hole.
35. Countersink the hole for a No. 6 flat head screw.
36. Sand or file the end radii.
37. Sand all surfaces.

Finishing and assembling the game
Applying finish to the parts:
38. Stain four pegs a dark color and let dry.
39. Apply a surface finish to the board, peg storage hole cover, and pegs.
40. Allow all finishes to dry properly.
Assembling the product:
41. Place the dark pegs in one storage hole and the light pegs in the other hole.
42. Make the screw hole for the peg storage cover. It should be over the anchor hole in the game board.
43. Attach the cover with a 1/4 in. by No. 6 flat head wood screw.

CHALLENGING YOUR LEARNING

Make a chart like the one in Fig. 2E. List the steps in the procedure where you used each type of tool:

TYPE OF TOOL	PROCEDURE STEP
Measuring Tool	
Cutting Tool	
Drilling Tool	
Gripping Tool	
Pounding Tool	
Polishing Tool	

Fig. 2E. List where you used each tool.

Chapter 3
Materials and Technology

materials to make things? Do you know the names of the four major types of materials used to make products? How about the difference between a material which provides energy and one that is used for its structure? (Structure means that a material has the ability to support certain kinds of activities. For example, wood can supply energy when it is burned. But its "structure" makes it strong. Thus, it can be made into a bench. People can sit on it. The wood supports "sitting" activities because of its structure.)

We live in a world of **materials.** Everything around us is made of materials, Fig. 3-1. Some of these materials appear in nature. We call these natural materials.

Trees, grass, soil, minerals, petroleum, and many other materials are found naturally on earth, Fig. 3-2. Most of these materials have limited use for us. Trees give us shade on a hot day. Grass makes lawns and parks look better.

However, most objects around us are made of materials changed by people. The desks, chairs, and tables in your school were once natural materials. The table tops may be made of wood from trees. The plastic chair seats may have been made from natural gas.

How much do you know about materials? Do you know how people find, refine, and use

Fig. 3-1. We are surrounded by things made from materials.

Fig. 3-2. Natural materials include things that grow on or are found in the earth. Left. Trees grow in soil and provide wood. Right. Minerals are found underground.　(Weyerhaeuser Co. and Amoco Corp.)

This chapter will help you find answers to questions about materials. It will explore:
1. Types of materials.
2. Finding and processing materials.
3. Selecting materials for a job.

TYPES OF MATERIALS

Materials are substances out of which useful items are made. Three major types of materials are used in technological systems. One type provides energy to operate the systems. This type powers engines, turbines, and other energy converters. They may be petroleum products like fuel oil, gasoline, or diesel fuel. Falling water and wind are also sources of energy. So are uranium, wood, grain converted into ethyl alcohol, and coal. These materials will be discussed in Chapter 4, Energy and Technology.

The second type of materials includes liquids, gases, and nonrigid solids. They may be used to support life. The air we breathe and the water we drink are very important materials. Water and fertilizers promote plant growth on farms. This type of material also cools industrial processes, lubricates moving parts, and provides the raw materials for products and thousands of other uses.

A third type of material will be the main focus of this chapter. This type is sometimes called *industrial* or **engineering materials.** They have a rigid structure. This means their shape is hard to change. Most of us know these materials as solids. Engineering materials, as shown in Fig. 3-3, are the base for all products which have a set form.

Typical products made of engineering materials are automobiles, garbage bags, refrigerators, baseball bats, party dresses, and beverage bottles. These products hold their shape. They have structure. Soap powders, tooth paste, lipstick, cherry cola, and pizza are different. All of these products are made from materials without structure. They must be placed in or on containers to hold their shape.

Engineering materials, as seen in Fig. 3-4, include four major categories (groups):
1. Metals.
2. Ceramics.
3. Polymers.
4. Composites.

Each of these materials is widely used in industry and by individuals. Each is very different from the others.

Metals

Metals are inorganic materials. They were never living things. All metals have similar internal structures. Their molecules are arranged in a boxlike framework. These structures are

Fig. 3-3. Steel and lumber are examples of solid or engineering materials. (USX Corp. and Weyerhaeuser Co.)

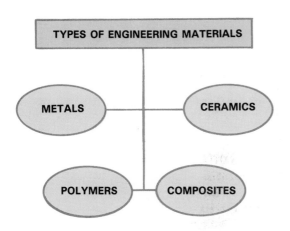

Fig. 3-4. There are four types of industrial materials.

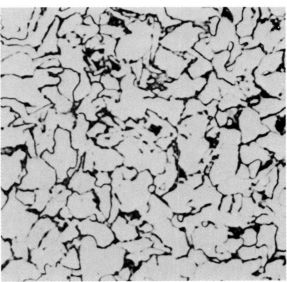

Fig. 3-5. This is what a section of metal looks like when magnified 1000 times. (Bethlehem Steel Corp.)

called crystals. They combine to form the grains of metal, Fig. 3-5. This type of framework produces a rigid, uniform material.

Metals are seldom used in their pure form. We do not often use pure gold, silver, copper, aluminum, or iron. Most metals we use are mixtures of two different base materials. Iron is combined with carbon and other elements to make steel. Aluminum is often combined with copper, silicone, and magnesium to make more useful materials. These mixtures of two or more elements are called alloys.

Daily, each of us comes into contact with metal alloys. Steel, brass, bronze, 14k gold, and stainless steel are all alloys.

Ceramics

Ceramics are probably the oldest material used by humans to make products. They are the materials of the Stone Age. Ceramics, like metals, are inorganic materials. Their structure is crystalline. They are made up of crystals.

Ceramic materials are very stable. Heat, moisture, or other environmental conditions do not affect them. Ceramic materials are stiff (do not bend) and brittle (break under stress).

There are four major types of ceramic materials. These are:

1. Clay-based materials: substances made up of crystals held in place by a glass matrix (binder). Typically, clay-based ceramics are used to make dinnerware (earthenware and china), sanitary ware (sinks, toilets, etc.), bricks, decorative tile, and drain tile.
2. Refractories: crystalline materials held together without binders. They withstand high temperatures. Fire brick and the space shuttle re-entry tiles are made of refractory materials.
3. Glass: amorphous (without a regular structural pattern) materials commonly made from silica sand. Glass is a very thick liquid which appears to be a solid under normal conditions. Glass is used for windows, containers, and many other products.

Polymers

Polymers are organic materials. The molecules form chainlike structures. They are produced from base materials which were once living. Most polymers are made from natural gas and petroleum. Wood and other cellulose fibers can also be changed into polymer materials.

Polymers may be natural or human made. A typical natural polymer is natural rubber. Synthetic (human-made) polymers are often called plastics. They are materials designed and produced to meet specific needs. For this reason, synthetic polymers are sometimes called engineered materials.

Polymers include three major groups. These are:

1. Thermosets: materials which form a rigid shape under heat and/or pressure. The structure remains rigid. Once set, it will not soften even when heat or water is applied.
2. Thermoplastics: materials which soften under heat and become hard (rigid) when cooled. They can be heated and reshaped a number of times.
3. Elastomers: materials which can be stretched but rapidly return to their original shape. To be called an elastomer, the material must withstand being stretched at least twice its length.

Composites

Composites are a combination of materials. However, each material retains its original properties. When mixed, cement, sand, gravel, and water form a composite. We call it concrete. The cement binds the sand and gravel together. Water starts the curing action of the cement. However, the sand and gravel have not changed. You can break up the concrete and remove the gravel. The sand is also there. However, it is hard to separate from the cement.

Likewise, wood chips are glued together to form particleboard. Thin wood sheets (veneer) are glued together to form plywood. And, wood fibers are heated and pressed. The natural glue (lignin) binds the fibers into a rigid form we call hardboard. These are all composites.

Still another common composite is made of glass fibers and plastic binders. The fibers are coated with a liquid plastic. When the plastic cures we have a product called fiberglass. It is widely used for automobile, boat, and truck parts; bathroom shower enclosures; cafeteria trays; and sporting goods.

GETTING MATERIAL RESOURCES

All materials we use can be traced back to the earth. As you have learned, they may be living things. Or they may be minerals and other elements found in the ground, sea, or air. These materials may be grouped as:

1. Renewable material resources.
2. Exhaustible material resources.

Renewable Resources

Renewable resources are living things. They are born or sprout. Then they grow and mature. Finally they die, Fig. 3-6. These materials may go through their life-cycle without human care. They may be part of what

Fig. 3-6. Young trees grow next to a stand of mature trees. Why should mature trees be cut rather than younger ones? (Weyerhaeuser Co.)

we call "nature." But many of these materials are grown by people. Humans engaged in *farming* (growing food and fibers) and *forestry* (growing trees) may produce these materials.

Commonly used renewable material resources are trees, grains, fish, animals, fruits, and vegetables.

Locating and harvesting

Some renewable materials must be located before they can be harvested. These are the materials growing wild in nature.

Foresters search the forests to locate the mature trees. They also select the correct specie, Fig. 3-7. Some trees have value for lumber, others for paper. Still others are good only for firewood. In fact, some trees have little or no commercial value. The western juniper and the pin oak are two examples. You may want them in a yard for shade and appearance. However, they produce little valuable lumber or wood fibers.

Fish also must be found before they are harvested. Many people will tell you that finding a tree is easier than finding fish.

Growing and harvesting

Most people are familiar with farming. Farmers plant seeds in the spring. They care

for the crop during the growing season. Weeds and pests are controlled. Fertilizer applied to the soil increases growth. At the end of the growing season the crop is harvested, Fig. 3-8.

In a similar manner, ranchers raise cattle and sheep. Adult livestock give birth to their young each year. The young animals are fed and cared for. They grow to be young adults. Some are saved to raise additional young animals. The majority are "harvested." They are butchered for their meat, hides, and other parts.

Fig. 3-7. Foresters carefully select the trees that will be harvested. (Weyerhaeuser Co.)

Fig. 3-8. Farmers harvest mature crops, like this grain, before they spoil. (Deere and Co.)

Materials and Technology   51

But farming and ranching are not the only ways renewable resources are grown by people. Many wood products companies operate large tree nurseries. They grow trees to be planted in forests or on tree farms. The newly planted trees are carefully tended. They receive fertilizer and often are pruned to increase growth. The goal is to grow the maximum amount of wood fiber in the least time.

Similarly, fish are now being raised on farms. A large industry has emerged to grow catfish for sale.

The key for all renewable resources is to harvest them when they are *ripe* (ready to be used), Fig. 3-9. Cutting trees which are too small is wasteful. Also, letting a tree mature and die wastes valuable resources.

Fig. 3-10. Modern farming practices help protect soil from erosion. (Deere and Co.)

Fig. 3-9. This machine harvests mature trees. They will be made into paper.

Likewise catching young fish breaks nature's cycle. Later we pay a price. There will be too few mature fish to provide food.

Exhaustible Resources

Many resources are **exhaustible.** There is only so much of the resources on earth. When they are used they will be gone. This type of resource provides a unique challenge. People must use each resource very wisely. Wasted resources are gone forever. Changes in habits and attitudes will not bring them back. New resources cannot be grown.

Common exhaustible resources include petroleum, natural gas, *mineral ores* (iron, copper, aluminum, etc.), farm soil, clay, and *chemical deposits* (salt, sulfur, etc.), Fig. 3-10.

Obtaining exhaustible materials involves two major steps. These are:
1. Finding deposits of the resource.
2. Extracting the raw material.

Finding the resource is often the job of geologists, Fig. 3-11. They study the earth to determine possible locations of the materials. They may use aerial maps and satellite pictures. Seismic studies are also used. These involve firing explosives buried in the ground. The resulting shock waves are "read" on special equipment. A "picture" of the rock layers below the surface is obtained. These "pictures" are charts and graphs made up by the equipment. They are read to locate promising sites for exploration.

Once a body of ores, minerals, or petroleum is located, it must be extracted. Holes are often drilled to get core samples. These are samples of the rock below the surface. The samples provide additional evidence of the presence of the resource. If the samples show promise that the resource is there, extraction begins.

Two major methods of extraction are used, Fig. 3-12. Wells are drilled to remove liquid resources. Mines are dug to remove solid resources. Some may be a big hole called an open pit mine. Others are tunnels into the earth called underground mines.

Fig. 3-11. This geologist searches for natural resources located in the earth. (Amoco Corp.)

EXTRACTING NATURAL RESOURCES

MINING

DRILLING

Fig. 3-12. Natural resources are extracted by mining and drilling. (Caterpillar Inc. and Amoco Corp.)

PROCESSING RAW MATERIALS

Raw material must be processed. It must be changed to make it useful. Often the resource found is mixed with other materials. In this condition it has limited value. Iron ore is of little use to us in its natural state. Likewise, a pile of harvested trees cannot be used easily.

Processing raw material falls into three general categories. These are:
1. Mechanical processing.
2. Thermal processing.
3. Chemical processing.

Mechanical Processing

Mechanical processing uses machines to cut, crush, or grind the material into a new form. Grain is ground into flour. Rocks are crushed into gravel. Trees are cut into lumber, veneer, and chips, Fig. 3-13.

Fig. 3-13. Saws cut logs into lumber. Sawing is a mechanical means for changing material. (Weyerhaeuser Co.)

Thermal Processing

Often we rely on heat to process a material. This is called **thermal processing.** Iron ore, coke, and limestone are heated together in a blast furnace. Pig iron is produced. Later this material is heat processed into steel.

Most metal ores are processed using heat. They are first crushed into manageable pieces. Heat is then used to melt the metal away from the impurities.

Chemical Processes

Many materials are processed using **chemical** action. Gold is removed from its ore using an arsenic chemical process. Natural gas changes into plastics when chemicals react with it.

Aluminum is produced using a combination of electricity and chemicals. This is called an electrochemical process.

The product of a processing action is an industrial material. This type of material comes in a standard form or size. It is called standard or industrial stock. Each of these materials needs further processing to be useful. Flour mixes with other materials to make bread. Lumber must be cut and shaped into useful products. This further processing is called manufacturing. The process will be studied in Chapter 8.

PROPERTIES OF MATERIALS

There are hundreds of materials available. They are produced to meet human needs and wants. For each job (product or use) a material must be selected, Fig. 3-14. It would be unwise to use wood for a heat shield. Even a low temperature would cause it to burn. Steel would be a poor choice for a baseball bat. It is too *heavy* (dense).

Understanding **material properties** helps people select the right material for the job. *A material property is a characteristic the material has.* These properties can be divided into seven groups:
1. Physical.
2. Mechanical.
3. Chemical.
4. Thermal.
5. Electrical and magnetic.
6. Optical.
7. Acoustical.

Fig. 3-14. Each material used to build and equip this kitchen was chosen to meet special requirements. (Coachman Industries)

Physical Properties

Physical properties describe the basic features of a material. One of these features is density. This is a comparison of weight to size. Density is often given in pounds per square foot.

Another physical property is moisture content. It tells us the amount of water in the material. Lumber used for furniture has a moisture content of six to eight percent. Construction lumber will have a moisture content in the 12 to 16 percent range.

A third physical property is surface smoothness. This property will affect a material's use. A very smooth material makes a poor floor. It becomes too slippery when wet. Likewise a rough material is a poor bearing surface. The parts will not slide easily. Abrasive materials have a rough surface while glass has a smooth surface.

Mechanical Properties

Mechanical properties affect how a material reacts to mechanical force and loads, Fig. 3-15.

They describe how a material will react to twisting, squeezing, and pulling forces.

Strength is a common mechanical property. It measures the amount of force a material can take before breaking. A strong material can

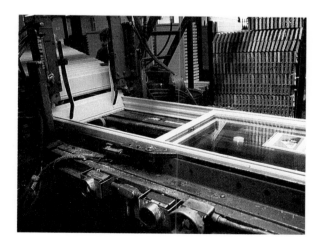

Fig. 3-15. These windows are manufactured to withstand many mechanical forces. (Andersen Corp.)

Materials and Technology 55

hold heavier loads than weak materials. Steel is very strong while writing paper is fairly weak.

Hardness is another mechanical property. This property measures the resistance to denting and scratching. Often, hard materials are also brittle. They break easier than softer materials. Diamonds and glass are both hard materials. They will *shatter* (break) if they receive a sharp blow.

A third mechanical property is ductility. This measures the ability of a material to be shaped with force. A ductile material can be hammered or rolled into a new shape. Many aluminum and copper alloys are very ductile. They are easy to form into complex shapes.

Chemical Properties

Chemical properties describe a material's reaction to chemicals. These chemicals may be of any kind. They may be water, some other liquid, or acids in the air.

These chemicals may cause corrosion. They may react with the material causing unwanted byproducts. (Water reacts with steel to form rust. Copper reacts with chemicals to tarnish.)

However most ceramic materials resist chemical actions. That is why ceramics are used for dishes, bathroom fixtures, and food containers.

Thermal Properties

Thermal properties relate to a material's reaction to heat. Most materials *expand* (become longer and wider) when heated. This property is called thermal expansion. The property can be used in many ways. For example, gently heating a metal jar lid may make it easier to remove. The metal lid expands more than the glass jar.

Many materials allow heat to move through them. The measure of this property is called thermal conductivity. The higher the conductivity, the easier heat will move through the material. If you heat one end of a metal rod the other end will quickly become hot too. Most metals have high thermal conductivity. But wood and most plastics will catch fire before

the other end gets warm. They have low thermal conductivity. This explains why most kettles are made of metal while their handles are plastic, Fig. 3-16. The metal conducts the heat quickly to the food. The handles insulate the heat from the cook's hands.

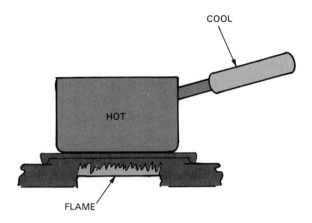

Fig. 3-16. A material's thermal properties can be used to advantage when designing a product.

Electrical and Magnetic Properties

These properties describe the material's reaction to electrical current. Some material will easily carry the current. They are called conductors. Others resist the flow of electricity. They are called insulators. Metals are generally good conductors. Most polymers and ceramics are good insulators. Plastic insulated copper wire is used to carry electricity in homes and other buildings.

The same is true of magnetic forces. Some materials will become magnetized easily. Other materials will never become magnets. Most iron alloys (steel) can be magnetized. Almost all other materials are nonmagnetic.

Optical Properties

Optical properties govern the materials reaction to light. Some materials stop all light. They are said to be opaque. Others let light pass through them easily. They are called transparent.

Another optical property relates to reflecting light. Each material absorbs part of the light striking it. Other parts of the light spectrum are reflected. The part that is reflected reacts with the human eye. We see the material as having color.

Acoustical Properties

This property determines the material's reaction to sound waves. Some materials absorb the waves. They are sound insulators or deadeners. Others reflect sound. Some materials will carry sound. They are sound transmitters. Smooth hard materials generally reflect sound. Softer, uneven material will absorb sound.

SUMMING UP

Materials are key to our existence. They are used for clothing and shelters. They become products and fuels. They are also food and fibers.

An important group of materials are called engineering materials. They have a set structure. They are the rigid materials used in many durable products.

Engineering materials are the metals, ceramics, polymers, and composites. Each type of material has its own use and set of properties. These properties include the physical, mechanical, thermal, chemical, electrical and magnetic, optical, and acoustical.

All materials are either renewable or exhaustible. Renewable resources can be grown and harvested. Proper management will give us a constant supply of these resources. Exhaustible resources have a limited supply. They are the minerals and other materials of the earth. Once used they cannot be replaced. They require careful use if future generations are to have them too.

All materials must be grown or located. They then must be harvested or extracted. Finally, they are processed into industrial or standard materials. These materials are the imputs to manufacturing, construction, communication, and transportation systems.

KEY WORDS

These are words used in this chapter. Do you know their meaning?

Acoustical properties, Ceramic, Chemical processing, Chemical properties, Composite, Ductility, Electrical and magnetic properties, Engineering materials, Exhaustible resource, Material, Mechanical processing, Mechanical properties, Metal, Optical properties, Physical properties, Polymer, Processing, Property, Renewable resource, Thermal processing, Thermal properties.

ACTIVITIES

1. Look around the room and identify five uses of (a) metals, (b) ceramics, (c) plastics, and (d) composites.
2. Complete an experiment which compares a property, such as strength, for two different materials.
3. Obtain a material kit from your teacher. Arrange the materials in the kit first by hardness, then by smoothness, and finally by density.

TEST YOUR KNOWLEDGE
Chapter 3

Do not write in this text. Place answers to test questions on a separate sheet.
1. Indicate which of the following materials are natural:
   a. Plastic.
   b. Plants.
   c. Rocks.
   d. Lumber.
   e. All of the above.
   f. None of the above.
2. Give a definition of natural material.

3. Read the description and name each type of material:
  a. Power engines, turbines and other types of converters: _____
  b. Provide support for life: _____
  c. Have a rigid structure: _____

4. Indicate which of the following are engineering materials:
  a. Butter.
  b. Metal.
  c. Glass.
  d. Plastics.
  e. Concrete.
  f. Petroleum.
  g. All of the above.

5. Living things are called _____ resources. Those which cannot be replaced, once used are called _____ resources.

6. Finding buried natural resources is often the job of _____.

Match the processes at left with the description at the right.

7. ____Mechanical processing.   a. Producing plastic from petroleum.

8. ____Chemical processing.   b. Producing steel.

9. ____Thermal processing.   c. Making lumber.

10. List seven properties of materials and define each.

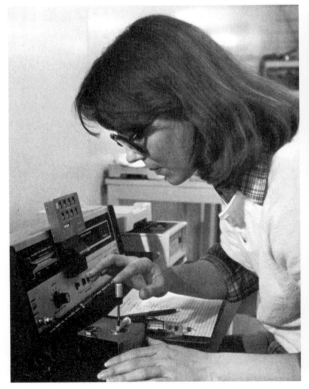

Earlier in this chapter you read that materials are picked for products because they have certain properties. After manufacture, the new product must be tested. They must meet set standards for strength and other properties. Left. A test machine measures the coating of platinum on an oxygen sensor (measures in engine exhaust gas oxygen). Right. A burst test is being made on another auto part.   (AC Spark Plug—GM)

APPLYING YOUR KNOWLEDGE

Introduction

The world is made of materials. Science tells us that materials are found in three states: gases, liquids, and solids. Solid materials are often called industrial materials. They can be grouped as metals, ceramics, polymers, and composites.

Each material has a unique set of properties. These properties tell us how a material will act or perform under certain conditions. As discussed in this chapter, the seven major material properties are:

1. Physical.
2. Mechanical.
3. Chemical.
4. Thermal.
5. Electrical and magnetic.
6. Optical.
7. Acoustical.

In this activity you will be able to test your knowledge of materials and their properties.

Equipment and Supplies

Material kit
Data recording sheet (sample provided, See Fig. 3A below).
Metalworking vise
Pliers
Ball peen hammer
Mill file
Flashlight

SAFETY: Review with your instructor any safety rules concerning use of tools. If not properly handled they may cause injuries.

Procedure

Your teacher will divide the class into groups of three to five students. Each group should:
  1. Obtain a material kit. It will contain a set

DATA RECORDING SHEET MATERIAL IDENTIFICATION AND PROPERTY TESTING						
Group: _____ Date: _____						
Specimen number	**1**	**2**	**3**	**4**	**5**	**6**
Name of material						
Type of material						
Density						
Hardness						
Ductility						
Light Reflectivity						
Torsion Strength						

Fig. 3A. Sample data recording sheet. (Do not write in the book.) Fill in the sheet given you as materials are tested. On the back of the sheet, list products for which each material is suited. Remember, manufacturers select a material for its properties.

of six materials. You will need two samples of each material. Samples should be the same size. They should also be numbered.

2. Obtain a data recording sheet.
3. Carefully look at each material. Write down the name of each material (such as, copper, wood, steel) on the chart.
4. List under its name whether the material is a metal, ceramic, polymer, or composite.
5. Select one sample of each material.
6. Arrange the samples by weight. Since they are the same size this will show which are denser.
7. Put a "1" in the "density" column for the heaviest sample, a "2" for the next heaviest, etc.
8. Test the hardness of each sample. Take two file strokes on the surface of each sample, Fig. 3B. The harder the material the less the file will cut.

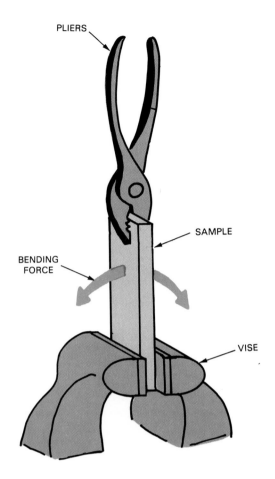

Fig. 3C. Testing the ductility (ability to bend) of a material.

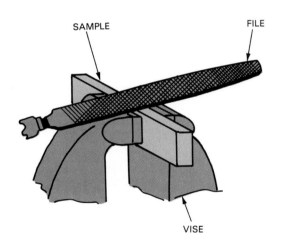

Fig. 3B. How to test a material for hardness.

9. Mark a "1" in the box of the hardest material, a "2" for the next hardest, etc.
10. Test the ductility of each sample. Place one end of the sample in the vise. Grip the other end of the sample with a pair of pliers. Bend the sample back and forth three times, Fig. 3C. Note how easy it is to bend each sample. Also look for any cracking or breaking along the bend line.

11. Place a "1" in the column for the material which (a) was easiest to bend and (b) did not crack. This is the most ductile sample. Place a "2" for the next most ductile sample, etc.
12. Take the second set of material samples.
13. Test the light reflecting property of the material, Fig. 3D. Shine the flashlight on each sample. A material which "shines" is reflecting light.
14. Place a "1" in the column of the material which reflects the most light. Place a "2" for the next best reflector, etc.
15. Test the torsion strength of each material. Place one end of the sample in the vise, Fig. 3E. Grip the other end of the material

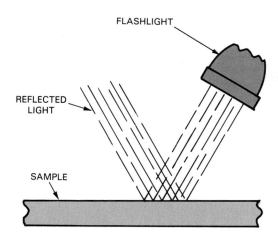

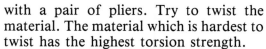

with a pair of pliers. Try to twist the material. The material which is hardest to twist has the highest torsion strength.

16. Place a "1" in the column for the material with the best torsion strength, a "2" for the next best, etc.

Fig. 3D. Use a flashlight to test ability to reflect light.

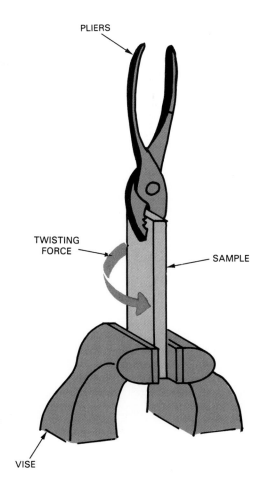

Fig. 3E. Making a torsion test.

Chapter 4
Energy and Technology

There can be no industry, no technology without energy. Energy comes from nature. We see its effects around us.

Wind blows. It pushes on sails and turns wind mills. Fuels burn and create heat. Rivers carry boats downstream. We see the effects of lightning and electricity.

But what is energy? When someone is very active we say that person has "a lot of energy." See Fig. 4-1. When tired, we have little or no energy.

An automobile moving down the street is using energy. When machines in a factory produce a product, energy is needed. When you turn on a switch, energy lights up your room.

Energy provides a force. *It is the ability to do work.* Work is the use of force to create movement.

Power and energy are not the same thing. Power is the amount of work done in a period of time.

Suppose that you pedal your bicycle one mile. That is work. But if you travel a mile in five minutes, that is power.

Power can be calculated with a simple formula. It is: Power = distance × force/time.

Fig. 4-1. Human activity requires energy. We get energy from the food we eat. It is our "fuel." (Kevin Smith)

HUMAN ENERGY

Primitive humans knew little of energy. Therefore, they did not develop technology. Their energy came from nuts, fruits, and the flesh of animals. Only human muscle was available to do work. Fig. 4-2 shows one way muscle power was used.

THE NATURE OF ENERGY

Energy can be changed from one form to another. See Fig. 4-3. When this happens, some of the energy is wasted. That does not mean it disappears. Energy is never destroyed. Still, it can become too weak to do any work. Remember how you feel after a lot of physical activity?

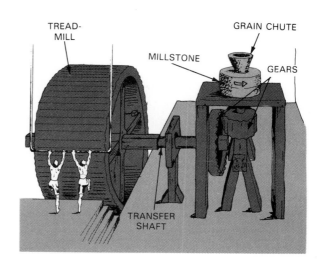

Fig. 4-2. Without machines to run mills, early humans had to use their muscles.

	CHEMICAL	ELECTRICAL	HEAT	LIGHT	MECHANICAL	ACOUSTICAL (SOUND)
Chemical	Foods Plants	Battery Fuel cell	Fire Food Hot water boiler Steam boiler	Lamp Gas lantern Candle Firefly	Gas engine Human muscle Animal muscle	Smoke alarm Exploding matter

Electrical	Batteries
	Electrolytes
		Diodes
		Transistors
		Transformers
			Electric
			  blanket
			Hairdryer
			Toaster
				Light bulb
				TV screen
				Lightning
					Electric motor
					Relay (type of
					  magnet)
						Horn
						Loudspeaker
						Thunder

Heat	Distilling
	Vaporizing
	Gasifying
		Thermocouple
		Thermopile
			Heat exchanger
			Heat pump
			Solar panel
				Fire
				Light bulb
					External and
					  internal com-
					  bustion engines
					Turbine
						Explosions
						Flame tube

Light	Camera film
	Plant growth
	  through sun-
	  light
		Photovoltaic
		  (solar) cell
		Photoelectric
		  cell
			Heat lamp
			Laser
				Laser
				Reflector
					Photoelectric
					  door opener
						Sound track of
						  movie
						Video disc

Mechanical	Gunpowder	Alternator Generator	Brake Friction	Flint Spark	Flywheel Pendulum Water stored in a tower	Wind instrument Voice

Acoustical (sound)	Hearing	Hearing
		Microphone
		Telephone
			Sound
			  absorption
				Color organ	Ultrasonic
					  cleaner
						Megaphone

Fig. 4-3. Energy can be changed from one form to another. The chart will show you what is used to make each kind of change. For example, to change electrical energy to heat, you could use a hair dryer.

Kinds of Energy

Energy is either at work or at rest. Have you heard the term **potential energy?** This is energy that sits around. It is waiting, but always ready to do work. This type of energy is found in the huge weight on top of a pile driver. See Fig. 4-4. As long as the weight sits there, it has only the potential (future ability) of doing work. When the weight is released, it drops with great force. It becomes **kinetic energy.** Its weight and force push the piling into the ground.

Another energy device is the mousetrap. To set it, you force the bail (U-shaped piece) from one end of the trap to the other. The spring tries to push the bail back to its original position. However, it is stopped by other parts of the trap. What do we have? Potential energy! Now, when something springs the trap, the bail snaps back to its original position. That's kinetic energy. It is the energy of motion.

The potential energy in the pile driver is not the same as the energy in the mousetrap. In the pile driver, the force comes from the weight's position above the piling. The force of gravity causes its energy. In the mousetrap the energy is in the spring. In setting it, you deform it (force out of its normal shape). This gives it energy. The spring wants to return to its original shape. We call this potential energy of deformation. Kinetic energy does the work. Still, potential energy is important too. It allows us to store energy.

Forms of Energy

Energy has different forms:

1. Radiant or light energy (atomic motion present in sunlight, fire, and any matter).
2. Heat energy (increased activity of molecules found in matter whose temperature has been raised).
3. Mechanical energy (movement produced by humans, animals, and machinery).
4. Chemical energy (reaction between substances such as petroleum and oxygen).
5. Electrical energy (movement of electrons in matter caused by lightning, batteries, and generators).
6. Nuclear energy (the energy from the splitting of atoms).

Fig. 4-4. Top. A pile driver drops a heavy weight to drive posts into the ground. The weight has potential energy when it is high above the post. As it is dropping it has kinetic energy. Can you explain why? Bottom. A tree standing in a forest has potential energy until it is felled by the lumberjack. (Weyerhaeuser Co.)

The Sun, Primary Energy Source

The sun is a huge energy "factory." Indirectly, it supplies 95 percent of the energy found on earth. Solar activity accounts for wind power, water power, and fossil fuels. See Fig. 4-5.

Fig. 4-5. Our sun supplies nearly all of the energy found on earth. The temperature of its surface is 10,000 °F.

Every day the sun gives the earth enormous amounts of heat and light. If all of it could be collected, two days of sunlight could provide more energy than all of the world's energy supplies. Even so, only 13 percent of all solar energy leaving the sun actually reaches earth.

How the sun works

Some of the sunshine creates kinetic energy. Another large share of it gets stored as potential energy.

Winds, a type of kinetic energy, come from uneven heating of the atmosphere by the sun. Heated air rises and causes cold, heavier air to move in under it. Winds are created. They are often strong enough to drive mills and electric generators.

The sun also works like a giant water pump. Radiant energy heats water until it vaporizes. These water vapors rise into the sky. Here they form clouds. See Fig. 4-6. When cooled, the vapors become a liquid again. The water falls as rain.

Fig. 4-6. The sun is a huge water pump. It draws water vapor into the atmosphere where it forms clouds. The clouds produce rain. (Natural Gas Supply Assoc.)

Much of the solar energy becomes stored in plant life. This energy remains stored until something releases it. Animals and humans eat plants for energy and growth. Burning releases the energy in wood and other plants.

ENERGY SOURCES

Technology needs energy that is plentiful and easy to control. When the energy is controlled, it can deliver power when and wherever needed.

For our ancestors energy sources were limited. People relied upon natural sources. They burned wood to cook food and keep warm. They had water and wind do work for them.

Still, these energy sources sometimes failed. During droughts rivers were low. Water wheels did not have enough water power to drive them. When winds did not blow, mills stopped working. Sailing ships couldn't move. The Romans had used so much wood that it was becoming scarce. They were starting to capture solar energy. They used it to heat their homes.

Today's Energy Sources

The term, "supplies," is often used to mean the energy sources we employ. They are ready to use anytime we need them. We usually store them where we can use them as needed.

Some energy sources are not used. There are good reasons why we do not or cannot:
1. They are scarce.
2. They are hard to collect.
3. They are hard to move from where they are to where they can be used.
4. They are not always available for use.

COMMON ENERGY SUPPLIES

Developed countries like the United States and Canada use several different types of energy supplies. The most-used and practical of these are:
1. Fossil fuels.
2. Electricity.
3. Nuclear energy.

Fossil Fuels

Fossil fuels are the remains of once-living matter. Dead plants or animals decay. This process of decay is also called **oxidation.** The decaying matter reacts with oxygen.

Fossil fuels are formed under conditions where there is very little oxygen, Fig. 4-7. The dead matter cannot fully decay. Some, but not all, of the carbon is released. As a result, the partially decayed matter becomes a deposit of material high in carbon. The material can be found in several forms. These include peat, coal, petroleum, and natural gas.

The carbon in fossil fuels makes them burn easily. Burning is rapid oxidation. It requires large quantities of oxygen.

Burning produces large amounts of heat. This is one reason why fossil fuels are popular. They have been a major source of energy for several hundred years.

Peat is decayed plant matter that sank in swamps. It is made up of moss, reeds, and trees. Other plant matter and soil pressed on it. The decaying matter compressed into solid matter. It became somewhat like coal. In some parts of the world it heats homes.

Coal is also decayed plant matter. It started out just like peat. However, it was compacted (made more solid) by the weight of soil and rock.

Coal was once widely used. It provided heat for powering steam locomotives. The locomotives pulled trains. Coal-fired stoves and furnaces were common in homes, factories, and public buildings. See Fig. 4-8.

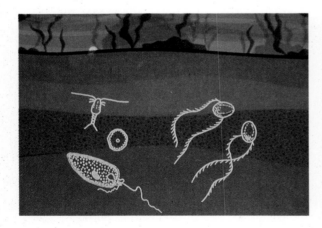

Fig. 4-7. Long ago many life forms flourished on earth. They absorbed energy from the sun and lived for a time. When they died they sank into the earth. These life forms became our fossil fuels.   (Standard Oil Co. of California)

Fig. 4-8. Coal is stored carbon. It is changed to heat energy by burning.
(Quaker State Oil Refining Corp.)

Gradually coal was replaced by petroleum products and natural gas. Petroleum is thought to be the decayed remains of plant and animal life. Like coal, it became buried and partially decayed. It is a liquid that has been pumped from pockets or "wells" deep below ground.

Petroleum requires **refining**. This is a heat process that separates the petroleum into more usable products. The products include liquids such as gasoline, diesel fuel, kerosene, and fuel oil. See Figs. 4-9 and 4-10. We depend on petroleum fuels for:
1. Heating buildings.
2. Powering transportation vehicles.
3. Operating manufacturing processes.

Natural gas is another form of petroleum. It is a gas instead of a liquid. Natural gas is used for most of the same purposes as

Fig. 4-10. A refinery worker installs insulation to refinery pipes. (Atlantic-Richfield Co.)

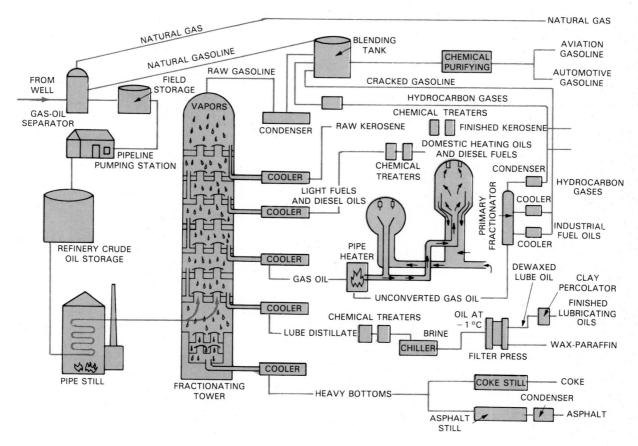

Fig. 4-9. When petroleum is refined many different fuels are produced. Vapors rise to different heights. This separates the different fuels.

petroleum. However, it is generally used for heating buildings. Sometimes gas is used to produce electric power or to create heat for manufacturing processes.

The process that produced fossil fuels is called **fossilization.** It still goes on. It continues at a low rate, however. One day the supply will be gone. We therefore call fossil fuels nonrenewable or exhaustible.

Water Power

Water power is energy resulting from the earth's gravity. Unlike fossil fuels it is a renewable energy. This means it will always be available. We know rivers flow from a higher level to a lower level. The water is constantly replaced by rainfall. Surface water is evaporated by the sun's heat. It falls back to the earth as snow, rain, and dew. This cycle occurs endlessly.

Early in time, humans learned how to use water wheels, Fig. 4-11. They were designed to catch the force of the moving water. The water

Fig. 4-11. This old waterwheel was designed to power a mill in Virginia.
(American Petroleum Institute)

turned the wheels, creating mechanical energy. A long shaft carried the energy to other wheels called pulleys. Belts or gears transferred the energy to machines.

Modern hydroelectric power stations use water stored behind dams (potential energy). The water falls through a turbine from great heights. The greater the fall the greater the force of the water. This height, from the water level to the turbine, is called the **head** or waterhead.

Nuclear Energy

Earlier you read that nuclear energy is the energy found in atoms. Scientists learned how to unlock this energy only recently. In the 1940s they discovered a way to split certain atoms of uranium. This caused other atoms to split. Each splitting is called a reaction. Once started the reactions continue. Each reaction causes more atoms to split. This is shown in Fig. 4-12. Also called **fission,** heat-creating reactions take place in a strong, carefully built structure called a nuclear reactor. The great heat produced is usually transferred to water. The water vaporizes into steam which spins a turbine. An electric generator coupled to the turbine, produces electric power.

More than 100 nuclear power stations operate today in the United States and Canada. See Fig. 4-13. Nuclear energy can also be harnessed for other uses. Several Navy ships are nuclear powered.

A second type of nuclear reaction causes parts of hydrogen atoms to *fuse* (join). This reaction is called nuclear **fusion.** Like fission, fusion releases huge amounts of energy. Our technology has been able to start fusion reactions. So far, it has been unable to keep them going.

Nuclear energy has some problems. It is difficult to build structures that will contain the energy produced by fission. Further, the nuclear wastes remain dangerous to life for many years. Proper storage is a major concern.

For this reason, many think that use of nuclear power should be discouraged. Others disagree. They have reason to believe that the wastes can be safely stored until their radioactivity is no longer a danger. See Fig. 4-14.

ALTERNATE ENERGY SOURCES

Alternate energy may replace or supplement (add to) nonrenewable supplies in the future. These sources are being studied. Efficient methods to harness and use them must be developed. Like water power, they will always

Fig. 4-13. Two power stations operate side by side. At left is a nuclear power station. The one at right burns coal.
(Carolina Power and Light Co.)

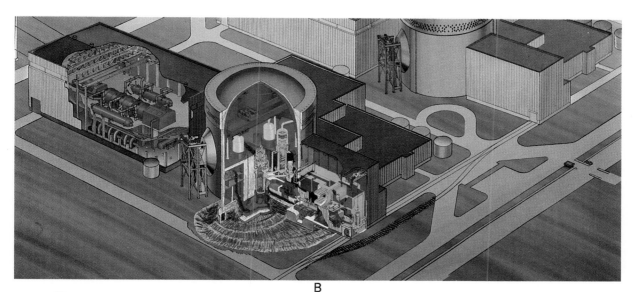

B

be available. We will never run out of them. These sources include:
1. Solar energy. This is direct use of the sun's rays. Solar cells can produce electricity directly from sunlight. Another use is having solar panels collect the sun's rays to heat water. The hot water can be used for household purposes or to heat homes. Some day solar furnaces may operate factories or power stations, Fig. 4-15.
2. Wind energy. (This is an indirect solar energy.) Before electric power became abundant, windmills dotted rural areas. Every farm had one. Now, wind generators that produce electricity are being developed and

Fig. 4-12. Nuclear energy. A—Splitting atoms of uranium creates heat. B—Cutaway of a nuclear power plant.
(Westinghouse Electric Corp.)

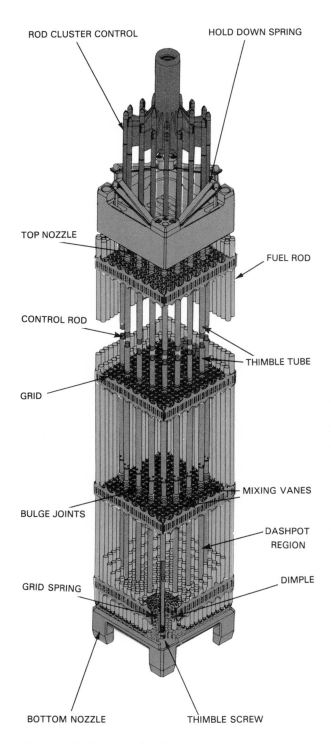

ROD CLUSTER CONTROL

HOLD DOWN SPRING

TOP NOZZLE

FUEL ROD

CONTROL ROD

THIMBLE TUBE

GRID

MIXING VANES

BULGE JOINTS

DASHPOT REGION

GRID SPRING

DIMPLE

BOTTOM NOZZLE

THIMBLE SCREW

Fig. 4-14. Nuclear fuel for a nuclear power station is contained in a fuel assembly like this. When the fuel is exhausted, the waste is stored so it cannot harm anyone. (Westinghouse)

A

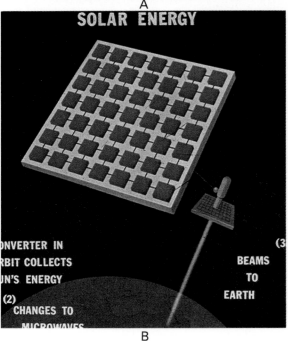

SOLAR ENERGY

CONVERTER IN ORBIT COLLECTS SUN'S ENERGY

(2) CHANGES TO MICROWAVES

(3) BEAMS TO EARTH

B

Fig. 4-15. A—The world's largest Solar Thermal Power plant uses large mirrors to create steam. Giant mirrors called heliostats, focus sunlight on a ''boiler.'' (Southern California Edison Co.) B—Artist's idea of a space power station. (American Petroleum Institute)

tested. See Fig. 4-16. The electricity can be used immediately, stored in batteries, or sold to power companies. As with solar energy, the power is expensive to produce. The source is not dependable. Sometimes there is no wind.

3. Tidal and wave energy, Fig. 4-17. Tides and waves have mechanical energy. The energy can be converted to a form we can use. Tidewater is trapped. Then, when it is released it creates electric power. Special generators can convert wave motion into electric power.

B

A

C

Fig. 4-16. A—This vertical axis turbine uses wind to generate electricity. It supplies electricity for 50 homes in Canada. (National Research Council, Canada) B—A horizontal axis wind turbine has a propeller like an airplane. C—Wind farm in Hawaii.

Fig. 4-17. Huge dams such as this can be built to catch and store water from high tides. Mouths of rivers that empty into oceans make good storage areas. The trapped water is released slowly to produce electric power. This one is at LaRance on the coast of France.
(American Petroleum Institute)

4. **Biomass** energy. Biomass includes all the living organisms in an area. Wood, crops, and wastes of plant, mineral, and animal matter are part of the biomass. Much of it is in garbage collected in cities and towns, Fig. 4-18. It can be burned for heat energy. It

Fig. 4-18. Garbage is a source of energy. It can be burned to produce heat. Mounds of rotting garbage produce methane gas (like natural gas).
(American Gas Assoc.)

can also be used to produce methane gas. There are also processes that can convert garbage into petroleum.

5. **Geothermal** and ocean thermal energy. Heated water from the ground and from oceans can produce energy. Deep beneath its surface, the earth is very hot, Fig. 4-19. Heated molten material is known as **magma**. It heats up ground water. Sometimes the heated water escapes to the surface through cracks in the earth, Fig. 4-20. Its energy can be used to run steam turbines and generators. Ocean thermal energy uses the heat stored in warm waters of oceans to generate electricity.

ENERGY SYSTEMS

A system is an organized way of getting things done. A lawnmower is one simple example of such a system. It is designed to cut grass. We think of it as a power system. A plant that produces electricity is also an energy system. See Fig. 4-21.

Conversion and Converters

An interesting thing about energy is that it easily converts from one form to another. In

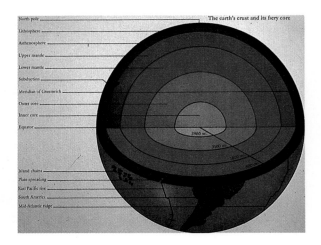

Fig. 4-19. Deep beneath its surface the earth is very hot. The inner core is so hot (above 3600 °F) that it is a fiery liquid.
(Union Oil Co. of California)

Fig. 4-20. Heat of the earth at some spots turns underground water to steam. Geothermal wells are drilled so the steam can be used. It can heat buildings or generate electricity. (Aminoil USA, Inc.)

fact, conversion is often necessary. Energy must be changed into more usable forms. Then it can be used to power machines. For example, burning fuel creates heat. The heat provides usable force. It delivers power where it is needed.

Energy will often be converted several times before it is finally used. Consider electricity. It may be produced by a fuel-powered generating station. Fuel, such as coal, is converted to heat energy. The heat energy of steam then drives a turbine. The spinning turbine (mechanical energy) turns an electrical generator. The electrical energy is carried over power lines to a user. There it may be converted back to mechanical energy by an electric motor. Some of it may be converted to light and heat energy by an electric bulb.

Many different kinds of devices are used for converting energy. In technology, machines are converters. However, animals or plants can be considered energy converters too. They take in food and change it to another form of energy.

ELEMENT	LAWN MOWER	POWER PLANT	EXPLANATION
Energy Source	Fuel and tank to contain it.	Piles of coal, natural gas pipeline, nuclear rods, force of falling water.	Gasoline is fed continuously to engine; power plants are supplied fuel on continuous basis—it is available night and day.
Conversion Method	Engine burns fuel and uses expanding gases created by combustion.	Fuels and fission create steam to rotate turbine. Falling water also spins turbine which operates electrical generator.	Gasoline engine converts chemical energy into heat energy so engine can create spinning motion (kinetic energy). Power station uses chemical, nuclear, or kinetic energy to create electrical energy.
Transmission Path	Piston and crankshaft	Electric power lines	Kinetic or mechanical energy travels from expanding gases to piston, to crankshaft to cutting blade. Electrical energy travels through wires to run electric motors and lights.
Control System	Throttle lever handle for steering	Switches and transformers	Mower's speed and direction are controlled. Transformers control amount of power coming through power lines. Switches turn electric power on and off.
Measuring Devices	Oil pressure gauge Gasoline tank gauge	Electric meters	In both systems, gauges tell operator operating condition of power system. Fuel, current, and voltage can be measured.
Load	Cutting blade	Electric motors Electric lights Electric heating	Load is the work done by systems: mower cuts grass; electric power runs machine, provides light.

Fig. 4-21. Big or small, all power systems have the same elements or parts.

Power to run technological devices is possible because of a variety of converters. Each type meets the needs of the technology for which it is used.

The Electrical Generator

Nearly one-fourth of all energy used in the United States and Canada is converted into electricity. This energy is then used in different ways. Some provides light. Another portion is used for heat. Some is used to power electric motors. Much of it is used to run various electrical and electronic devices.

Most electrical energy is provided by a machine called an electrical generator or **alternator,** Fig. 4-22. It converts mechanical (machine motion) into electrical energy. It is usually turned at high speed by a turbine or an engine. You have probably seen how a bicycle generator operates. It is powered by the rider.

Operation of a Generator/Alternator

Do you know how a generator works? It's principle of operation was discovered by an English physicist, Michael Faraday. He passed a copper wire through a magnetic field (an area of invisible magnetic force between the poles of a magnet). He learned that this caused electricity (charged particles) to flow through the wire. This is called **electromagnetic induction.**

The wire is part of a circuit (path along which electricity can flow). Electricity will flow through the wire. Imagine that you can see inside a very simple electric generator. It is a loop of wire turning in the air space between the poles of a magnet. See Fig. 4-23. As the loop spins, it moves through the magnetic field. This makes the electricity move in one direction through the wire. As the loop continues to rotate it becomes parallel to the magnetic field. The flow stops for a moment. The wires are not cutting through any lines of magnetic force. As the loop continues, current starts up again. Now it moves in the opposite direction through the wire loop. This type of electricity is called **alternating current.** Common electricity changes direction 120 times a second. It is called 60 hertz (cycles) alternating current.

Imagine now that an electric power station is simply a generator. All the homes and factories are part of the circuit. This is exactly how an electric power system works.

Fig. 4-22. A generator changes mechanical energy to electrical energy.

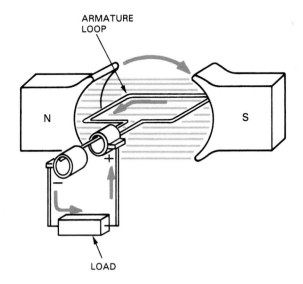

Fig. 4-23. A simple generator consists of a magnet and a loop of wire. The wire loop rotates through the magnetic field.

The heat engine

Another type of energy converter is called a **heat engine.** It burns fuels to create mechanical energy. Heat engines include:

1. Internal combustion engines.
   a. Two and four-cycle piston gas and diesel engines, Fig. 4-24.
   b. Jet engines.
   c. Gas turbines.
   d. Rockets.
2. External combustion engines.
   a. Steam engines.
   b. Sterling cycle engines.
   c. Free piston engines.

About one-third of all energy used is converted to mechanical energy in heat engines. They are used to move automobiles, airplanes, trucks, buses, ships, and other transportation vehicles.

Internal combustion engines

Internal combustion means that the fuel is first drawn into the engine. Then it is ignited and burned. The heated gases expand so rapidly they appear to explode. This provides the power needed to move the vehicle or cause some other type of motion.

Engines all operate much the same. A sealed chamber takes in a mixture of fuel and air. It burns there.

Fig. 4-25 shows the series of steps or stages for a typical piston engine. It has four steps or strokes:

Fig. 4-24. One of the best-known heat engines is the gasoline engine. (Buick)

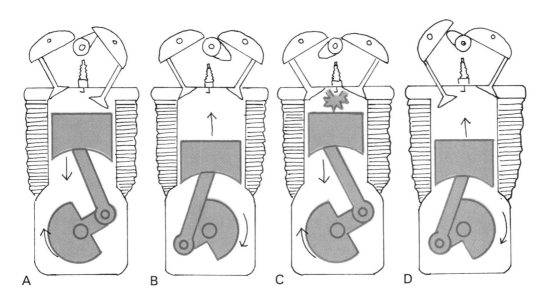

A   B   C   D

Fig. 4-25. How a four cycle engine operates. A—As piston moves downward, it pulls fuel into combustion chamber. (This is the space above the piston.) B—Piston squeezes fuel charge into a small space. C—Fuel charges burn. Hot gases force the piston downward. This is the power that does work. D—Burned gases are pushed out of the combustion chamber.

1. Taking in a fuel-air mixture (intake stroke).
2. Compressing the mixture (compression stroke).
3. Burning the fuel (power stroke).
4. Exhausting the gas (exhaust stroke).

This design is known as a four-cycle engine. Another type is the two-cycle engine. Unlike the four-cycle, it burns a fuel charge every time the piston is at the top of the cylinder.

Transferring engine power

You have learned that four-stroke internal combustion engines get power from exploding air-fuel mixtures. Look at Fig. 4-25 again. Expanding gases push on pistons. Rods connect the pistons to a crank. The rods transfer the motion of the pistons to the crank. When a piston moves the crank spins. Belts or shafts move the crank's force to turn wheels or operate other machines. See Fig. 4-26.

Jets, Turbines, and Rockets

Jet engines, gas turbines, and rockets, are other types of internal combustion engines which burn a mixture of fuel and air. The jet engine is used in airplanes. So are gas turbines. Rockets are used on spacecraft.

How a jet engine works

A jet engine is shaped like a tube. Both ends are open. Air and fuel mix and burn at the forward end of the engine. Compressed air enters the front of the engine. Fuel jets called injectors are located there too. As the air flows in, it is charged with fuel. The fuel sprays continuously. Spark plugs ignite the fuel-air mixture. Expanding gases escape out the back of the engine. Burning gas at the front of the engine is under great pressure. Gases at the back of the engine have little pressure. Unequal pressure provides forward thrust (force).

To understand this, think of the engine as an inflated balloon. As long as the neck of the balloon is tied, air cannot escape. The balloon has no motion. When the neck is open, air will rush out. There is no pressure at the back of the balloon. But pressure at the front of the balloon is still high. The unequal pressure (great at the front, less at the back) pushes the balloon through the air. See Fig. 4-27.

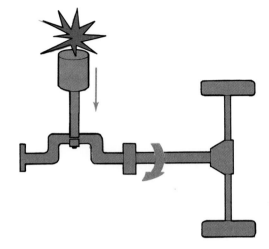

Fig. 4-26. This is how power is transmitted (moved) from an engine to the back wheels of an automobile.

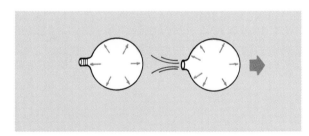

Fig. 4-27. A balloon will show us how a jet engine develops power. A—Air inside balloon pushes on all parts of the balloon equally. The balloon does not move. B—Air escapes from the balloon when the neck is open. Pressure is less on that end of the balloon. Pressure at the front causes the balloon to move in that direction.

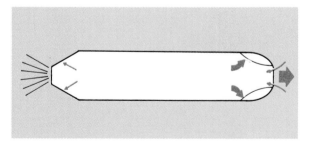

Fig. 4-28. Jet engine is like the balloon. Pressure is greatest opposite the exhaust. Jet moves away from the exhaust.

The jet engine's exhaust is like the air escaping from the balloon. It lowers the air pressure against the back end of the engine tube. But the pressure on the front end is high. It pushes the engine forward.

A rocket engine, Fig. 4-29, works something like the jet engine. It has one important difference. It carries its own supply of oxygen to mix with the rocket fuel. (Deep space has no oxygen.) A jet engine draws in air from the atmosphere.

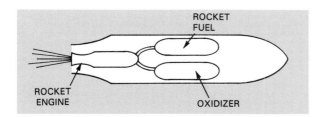

Fig. 4-29. This is how a rocket engine works. It is like a jet but carries its own oxygen supply so its fuel will burn in outer space.

External combustion engines

An external combustion engine burns fuel outside of the engine chambers. Only the expanding fluid enters the engine.

The steam engine is an example. It is used little today. Its operation and design is similar to the gasoline engine. Steam is introduced into the cylinder chamber. The steam moves the piston to turn a crankshaft. Spent (cooled) steam is exhausted from the engine and another charge of live steam is introduced as the cycle is repeated. See Fig. 4-30.

However, steam power is not out of date. Steam turbines produce almost 80 percent of electric power today.

The turbine has a wheel with blades on it. A jet of steam is aimed at the blades. The wheel spins like the crankshaft of a piston engine. The spinning power is tranferred to operate other mechanisms. Some turn propellers on ships. In power generating stations they turn huge electrical generators. Fig. 4-31 shows a simple drawing of a turbine.

Fig. 4-30. A steam engine pulled this oil drilling equipment near Independence, Kansas in the early 1920s. (American Petroleum Institute)

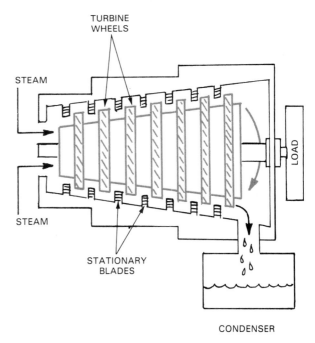

Fig. 4-31. Steam piston engine has been replaced by steam turbine. Steam pushes on blades of turbine. It spins and turns a generator.

Converters of the future

Scientists, physicists, and engineers are working on still newer types of energy converters.

These types will make direct conversions from one form of energy to another. This is desirable for several reasons. For one thing, direct conversions are more efficient. They do not waste as much of the energy. Also they should be more reliable. Direct converters do not break down as often. They usually have no moving parts.

The best known direct converter is the **fuel cell.** It looks like an electrical storage battery, Fig. 4-32. It can produce electricity from a fuel and oxygen. It does not burn the fuel.

A fuel cell has a container to hold chemicals such as phosphoric acid. The chemical is called an **electrolyte.** Two carbon plates are placed into the electrolyte. These are the terminals. Electricity will flow through them.

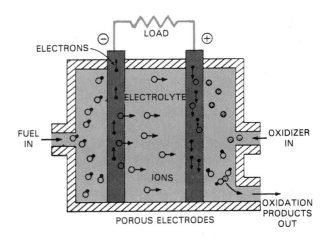

Fig. 4-32. The fuel cell makes electricity directly from fuel.

How a fuel cell works

To start the cell, oxygen and hydrogen (or another fuel) are fed into it. The fuel loses electrons to one of the carbon plates. The plate becomes negatively charged. Meanwhile, oxygen is fed to the other plate or terminal. The oxygen collects electrons from the second plate. This leaves the plate with a positive charge.

The fuel cell can now provide electric current. This current becomes active when a load, such as a light, is connected into the circuit. Free electrons on the negative plate travel through the conducting wire. They flow through the light bulb and back into the fuel cell through the other plate. In the electrolyte, ions of the hydrogen combine with oxygen ions to form water. (Water is a combination of hydrogen and oxygen atoms.)

Solar Cell

Another type of direct energy converter is the **solar cell,** Fig. 4-33. It makes electricity out of sunlight. A group of cells can be linked together. They have electrical currents (paths) built into them. When the sun shines on the cells there is a heat buildup. The heat drives electrons through the cell's circuits. The cells can be hooked to a load like a battery.

The solar cell still costs too much for the power it produces. However, it is being used

Fig. 4-33. This solar collector supplies both hot water and electricity. Photovoltaic cells are at top of center panel. Water is heated in three large panels.

where power lines cannot go. Solar cells provide power for space vehicles and telephones in remote areas.

Other direct converters of electricity are the **MHD generator,** and the thermoelectric generator. Like the solar collector panels, they operate on heat energy.

MHD stands for magnetohydrodynamic. Superheated gases are passed through a magnetic field. Electrons leave the gas and enter conductors in the generator. This causes an electric current.

A thermoelectric device generates electricity when heat is applied where two unlike metals are joined. Heat causes electrons to leave one of the metals and travel through the other. If the opposite ends of the two wires are joined, the electrons will flow through the wires.

The electrical energy produced by this type of converter is used to produce signals. We call such energy converters **transducers.** They change energy into information. This information can be used to control machines by sending them signals.

The thermoelectric device just described is used in heating systems. It signals that the pilot light is lit. Fig. 4-34 shows a small thermoelectric generator.

Fig. 4-34. A thermoelectric generator turns heat into electricity. Heat can drive the electrons through a circuit.

ENERGY AND POWER TRANSMISSION

Energy, or the power it produces, is not always used on the spot. Moving it to where it performs work is called **transmission.**

Shafts

Earlier, you learned that a car's driveshaft was one kind of transmission system. It carries mechanical power from the engine to the drive wheels.

Belts

Belts are another way of transmitting mechanical power. The system includes at least two pulleys. One pulley is on some source of power. The other one is on the load. The belt transmits the power from one pulley to the other. A factory machine operated by an electric motor might use a belt to transmit the mechanical power of the electric motor. The fan in a furnace is linked to an electric drive motor by a belt and pulley system.

Gears and chains

Gears transmit mechanical power with teeth that mesh with other gears. Sprockets are toothed wheels that are linked by a chain. A bicycle is a good example of this kind of transmission. The gears on sprockets and the spares in chain links eliminate slippage between the power source and the load, Fig. 4-35.

Fluids are sometimes used to transmit power. In hydraulic systems the fluid is oil or water. In pneumatic systems air is used. Both types are included in the term, fluid power. A common example is the braking system of a car.

Fig. 4-35. A bicycle chain moves energy from the pedals to the bicycle's rear wheel.

Stepping on a brake pedal causes a piston to put pressure on brake fluid (oily substance). The fluid then pushes on another piston at the brake pads or shoes on each wheel.

Transmitting electricity

Electricity is thought to be the movement of electrons of matter through certain materials called conductors. This is called the "electron theory." The flow of electrons through conductors (wires) produces the energy in electricity.

Electricity is easy to transmit. Lines carry it from the power plant to wherever electricity is to be used. The lines are conductors made of either copper or aluminum. See Fig. 4-36.

Electric current leaving the generating station is fed into transformers. These are devices which can increase or reduce the force of electric current. These devices boost the voltage from around 13,800 volts to as high as 700,000 volts. The current is then carried to the point of use. There, other transformers reduce voltage. Transformers that boost voltage upward are called **step-up transformers** while those which lower voltage are called **step-down transformers.**

Robots

Robots are examples of machines that use more than one source of power. They are used to take the place of a human in jobs that are boring or dangerous. Robots use electric, pneumatic, and hydraulic power transmission. The "brain" of the robot is the controller. It uses electricity and electronic devices to give the robot instructions on what to do. It receives these instructions from a human.

Another part of the robot is called the manipulator. It is like the human arm. It does the work. In general, a hydraulic ram may be used where strength or force is needed by the manipulator. Pneumatics (compressed air) could be used where strength is not needed.

USING AND CONSERVING ENERGY

The United States and Canada are among the most highly developed countries of the world. Our technological world uses a great deal of energy.

However, much of the energy taken from our environment is not used efficiently. It is not possible to stop all the losses of energy. But it is possible to make better use of the limited supplies. We call this conservation.

Conservation Measures

Energy and power technology have many benefits. But, there are some problems. Our technological society requires a great deal of energy. But most of the fuels we use are limited in supply. Most authorities agree that oil

Fig. 4-36. Power lines carry electricity from power stations.

resources will be used up sometime after the year 2000. Coal reserves may last 200-300 years.

Some effort is already being made to save energy:

1. Manufacturers are building more energy efficient products. Automobiles are being made smaller with more fuel efficient engines. Furnaces are more energy efficient. Until a few years ago efficiencies of 35 to 50 percent were common. New furnaces have efficiency ratings from 80 to 96 percent.

2. Buildings are better insulated. Higher insulation standards are being used in new construction. This means thicker insulation in ceilings and walls. Triple glazing (three panes of glass) is available in windows. A film on window glass keeps out summer heat. It keeps interior heat inside. Cracks are sealed better, cutting down on heat losses.

3. Industry is beginning to recycle more materials. This can be an important energy-saving step. For example, it takes large amounts of energy to make a ton of aluminum from bauxite ore. Converting scrap aluminum requires much less energy.

4. People have personal roles in conserving energy. We have formed habits that waste energy. Consider how we use electricity. How many times have you left radios and lights on when leaving a room? What would the savings be in energy if people car pooled to jobs or took public transportation instead of the family car? How many drivers observe the speed limit on highways and expressways?

Our standard of living is the highest in the world. We have many benefits which are the result of modern technology. Fig. 4-37 lists appliances and lighting found in the average home. We have a responsibility to do what we can to conserve natural resources for future generations.

Application: Energy Conservation

During the early 1970s there was an energy shortage in the United States. It was caused by an oil embargo which cut supplies of petroleum imports. Fuel for heating and transportation was scarce. In the years since then, federal regulations have dictated more fuel efficient automobile engines. Other energy converters are being designed to wring more energy out of the fuels or electricity they use.

People are being asked to help save energy. They are challenged to prevent waste. What are some of the things you and your family can do to conserve? What can be done in your school to use less energy? Do you think that our governments should do more about promoting energy conservation? Do you think government should pass laws forcing people to save energy? Why or why not?

SUMMING UP

Energy is the ability to do work. Work is the use of force to create movement. Power is work times time. Sometimes energy and power are confused.

Energy takes several forms. The forms include: light and heat energy, mechanical energy, chemical energy, electrical energy, and nuclear energy. Energy can be quite easily changed from one form to another. Still, it can never be destroyed. However, it can lose its ability to do work.

Energy can be at rest or stored. This is known as potential energy. Energy in motion is called kinetic energy. It is doing work.

Our energy comes from many sources. Most of them are the result of sunshine. Some of them we are able to use readily. Others are not easy to use.

If an energy source cannot be replaced, it is an exhaustible or nonrenewable source. If new supplies can be produced it is called renewable.

Our most-used energy source is the fossil fuels which are the partially decayed remains of plant and animal life. Other sources used are nuclear and hydroelectric energy.

We have still other energy sources which we call alternate sources. They are not now widely used. These include direct solar energy, wind energy, tidal and wave energy, biomass energy, and geothermal energy.

Energy is converted and used through an energy system. Such a system will have an energy source, a converter, a means of transmission, controls, a means of measurement, and a load. Much of our available energy is wasted. We need to conserve.

ENERGY CONSUMPTION BY HOME APPLIANCES AND LIGHTING		
	Annual Energy Consumption (kilowatt-hours)	Annual Cost of Energy Consumed*
Air Conditioner	2000	$ 123.00
Air Conditioner (room)	860	52.89
Electric Blanket	150	9.23
Can Opener	0.3	.02
Clock	17	1.05
Clothes Dryer	1200	73.80
Coffee Maker	100	6.15
Dishwasher (with heater)	350	21.53
Fan (Attic)	270	16.61
Fan (Circulating)	43	2.64
Fan (Furnace)	480	26.52
Fluorescent Light (3 fix)	260	15.99
Food Freezer (16 cu. ft.)	1200	73.80
Food Mixer	10	.62
Food Waste Disposer	30	1.85
Frying Pan	240	14.76
Hair Dryer	15	.92
Hot Plate (2 burner)	100	6.15
Iron (hand)	150	9.22
Light Bulbs	1870	115.01
Oven (microwave only)	190	11.69
Radio (solid state)	20	1.23
Radio Phonograph (solid state)	40	2.46
Range with oven	700	43.05
Range with self-cleaning oven	730	44.90
Refrigerator (frost-free) (12 cu. ft.)	750	46.13
Sewing Machine	10	.62
Shaver	0.6	.04
Television (black/white)	400	24.60
Television (color)	540	33.21
Toaster	40	2.46
Vacuum Cleaner	45	2.77
Washer (automatic)	100	6.15
Totals	12,911 Kwh	$794.07

*Cost of electricity = 6.15 cents per kilowatt-hour

Fig. 4-37. This is how we use energy in our homes. Do we waste some of it?

KEY WORDS

These words were used in this chapter. Do you know their meaning?

Alternating current, Alternator, Biomass, Chemical energy, Conductor, Conservation, Diesel fuel, Distillate, Electrical energy, Electrolyte, Electromagnetic induction, External combustion engine, Fission, Fossil fuels, Fossilization, Four-cycle engine, Fuel cell, Fusion, Geothermal, Gravity, Head, Heat energy, Heat engine, Internal combustion engine, Kinetic energy, Load, Mechanical energy, MHD generator, Nuclear energy, Oxidation, Potential energy, Radiant energy, Refining, Robot, Solar cell, Step-down transformer, Step-up transformer, Transducer, Transmission, Thermocouple, Work.

ACTIVITIES

1. Go to your school library and research the building of solar stills. Construct a simple still and experiment with muddy water. Distill the water by placing the still in direct sunlight. Explain to the class the principles behind solar distilling.
2. Visit an electrical power plant. Ask about the work and the types of jobs available. Write a report on what you heard and observed. Use a computer to prepare the report, if you can.
3. Build a small model of a water wheel. Demonstrate it to the class and explain how it works.
4. Make a list of devices you see in your community which use energy in any form. Place them in the order of their greatest importance to the well-being of the community.
5. Talk to your teacher about inviting someone from an energy company to speak at your school. Discuss with class members the topic or topics you would like the person to cover.

TEST YOUR KNOWLEDGE
Chapter 4

Do not write in this text. Place answers to test questions on a separate sheet.

1. Power is the ability to do work. True or false?
2. _____ energy is a term meaning that the energy is there but it is resting.
3. Tell whether the following situations are examples of potential energy or kinetic energy:
   a. A bat that is resting on a ballplayer's shoulder.
   b. A child sliding down a sliding board.
   c. A bat being swung by a ballplayer.
   d. An automobile rolling down a hill.
4. Explain how potential energy of gravity is different from potential energy of form.
5. _____ energy is the one causing motion in bodies.
6. As a source of energy, the sun (select all correct answers):
   a. Supplies 95 percent of the world's energy directly.
   b. Has a temperature of 10,000° F at its surface.
   c. Could never supply the amount of energy contained in fossil fuels.
   d. Sends out energy that travels through space at the speed of light.
7. The lighter a moving body, the more inertia is has. True or false?
8. Chemical energy is (indicate which, if any, of the following are true):
   a. Energy stored in molecules and atoms.
   b. Stored energy found in plants.
   c. The energy locked in fuels such as wood, petroleum, and natural gas.
   d. The energy stored in a battery.
   e. Energy stored behind a dam.
   f. All of the above.
   g. None of the above.
9. Electrical energy is caused by the movement of _____ particles.
10. Explain how nuclear energy is similar to chemical energy.

11. _____ are a type of kinetic or mechanical energy resulting from the uneven heating of the earth's atmosphere by the sun.

12. The heat of nuclear power comes from the _____ of atoms.

13. Without the earth's gravity, hydroelectric power would not be possible. True or false?

14. Define alternate energy and list five sources.

15. Which of the following are NOT part of a model energy system?
    a. An energy source.
    b. A person to use the energy.
    c. Something to convert the energy to a form that is usable by the system.
    d. Something to transfer the force from the converter to where the force is needed.
    e. A means of controlling what the power does.
    f. A way of measuring conditions important to the system.
    g. A load for the system to move or some condition that it must change.

16. An electrical generator is one type of energy _____.

17. Tell why conservation of energy is important.

All coal deposits had their beginning in swamps such as the Cole Creek Swamp in Mississippi. (National Park Service)

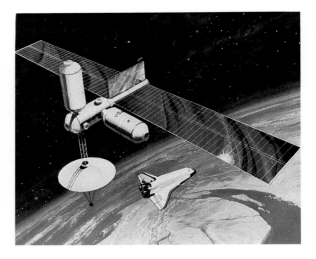

Collecting energy from space. Space experts are studying ways of developing power stations in space. Artist for Rockwell International has produced this drawing for a 50 kilowatt (50,000 watts) power module. It would beam energy (collected from sunshine) down to earth.   NASA

Falling water is a renewable energy resource. It will always be there.

APPLYING YOUR KNOWLEDGE

Introduction

People have always looked for ways to do work more easily. They wanted to do jobs efficiently and quickly. Therefore, humans have invented tools and machines. One of the most important inventions is the electric motor. We can find such motors in use almost everywhere. They power machines in factories. They run our refrigerators and freezers. They move air through furnaces and air conditioners. They move the hands on wall clocks. Think of your home. How many uses of motors can you list?

This activity will let you build a simple motor. With the help of your teacher you can see how the motor converts energy into power. As you can see in Fig. 4A, a motor changes electric energy into mechanical motion.

Device	Energy Input	Energy Output
Electric motor	Electric	Mechanical
Electric Generator	Mechanical	Electric
Battery	Chemical	Electric
Electric Light	Electric	Light and heat
Oil burner	Chemical	Heat and light
Wind mill	Mechanical (linear)	Mechanical (rotary)

Fig. 4A. Converting energy from one form to another. Can you add to the above list? Try to do this by listing five devices you see every day.

Equipment and Supplies

Material listed on the Bill of Materials, Fig. 4B.
72 in. of 18 gauge magnet wire
20 in. of 14 gauge solid uninsulated copper wire
Thread to tie the coil together
Coil winding jig (see Fig. 4C)
Bearing bending jig (see Fig. 4D)
1 1/2 or 6 volt battery
Scratch awl
Hammer or mallet
Wire cutters
Screwdriver
Utility knife
Abrasive paper

Part No.	Qty.	Description	Size	Material
1	1	Base	1/2 x 4 x 6	Plywood
2	2	Bearings	14 ga.	Copper wire
3	1	Coil	2 3/4" I.D.	Copper wire
4	4	Magnets	1/4 x 1" dia.	Speaker magnets
5	4	Screws	1/2 x 6	Sheet metal
6	2	Connecting wire	18 ga. x 12"	Copper wire

Fig. 4B. Bill of Materials for making an electric motor.

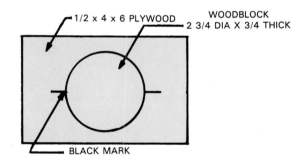

Fig. 4C. Coil winding jig. About five are required.

Procedure

Preparing to make the motor

1. Gather the materials needed to make the motor:
   a. 1 1/2" x 4" x 6" plywood base

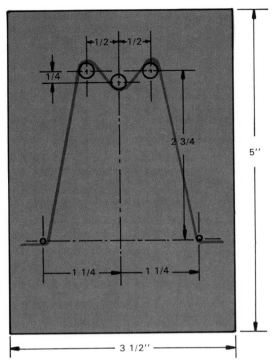

Fig. 4D. Bearing jig.

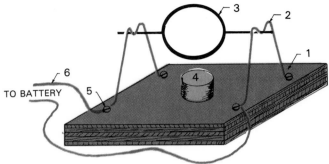

Fig. 4E. Electric motor. Refer to Bill of Materials for part names.

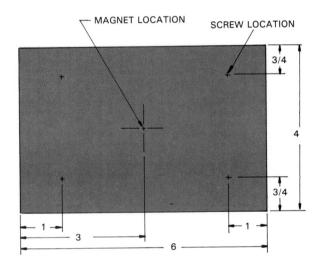

Fig. 4F. Motor base.

b. 1 pc. 18 gauge magnet wire, 96" long. (72" for the coil and 24" for the two connecting wires)

c. 1 pc. 14 gauge wire, 20" long

d. 12" thread

e. 4 - 1/2" x No. 6 pan head sheet metal screws

f. Four 1/4" x 1" magnets (or a substitute suggested by your teacher)

2. Each group of four to six students at a workbench should secure the following:

a. Coil winding jig.

b. Bearing bending jig.

c. Battery.

d. Scratch awl.

e. Hammer or mallet.

f. Screwdriver.

g. Utility knife.

Making the motor

Base (See Fig. 4F):

1. Lay out the location of the screw holes and the magnets.

2. Start the screw holes with the scratch awl.

3. Prepare for attaching the magnet. (Your teacher will show you how. The method will vary with the type of magnet used.)

4. Sand the edges and ends to remove sharp edges and splinters.

5. Start a screw in each of the four holes. Do NOT tighten the screw at this point.

Coil (See Fig. 4G):

1. Cut about 20 in. off the 18 gauge wire and lay it aside.

2. Mark 1 1/2 in. from one end of the long wire.

3. Carefully remove the insulation from the end of the wire up to the mark.

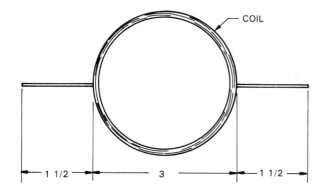

Fig. 4G. Completed coil for motor.

4. Place the wire in the coil winding jig. Leave the 1 1/2 in. clean end of the wire extending along one of the black marks.
5. Carefully wind the wire around the center core.
6. Stop winding the coil when you cannot make another turn and leave 1 1/2 in. on the opposite black mark.
7. Extend the wire along the black mark.
8. Cut the excess wire leaving 1 1/2 in. along the mark.
9. Remove the coil from the jig.
10. Tie thread at four places to hold the coil together.
11. Remove the insulation from the second end of the coil.

Bearing (See Fig. 4H):
1. Cut the bearing wire in half.
2. Use the bearing bending jig to form two bearings.
3. Cut off extra wire at the ends of the bearing.
4. Carefully remove the insulation from one foot of each bearing.

_____ **Assembling the motor**_____

1. Cut the 20 in. piece of 18 gauge wire in half.
2. Remove the insulation from each end of the wires.
3. Place one bearing under the screws at one end of the base.
4. Tighten the screw on the end which is still insulated.

5. Place the end of one of the connecting wires under the other screw.
6. Tighten the screw on the bearing and connecting wires.
7. Repeat steps 3 to 6 for the other bearing and connecting wires.
8. Attach the magnets.
9. Place the coil between the bearings.
10. Connect the motor to the battery.
11. Test the motor operation.
12. With the help of your teacher, describe how the motor converts electrical energy into mechanical motion.

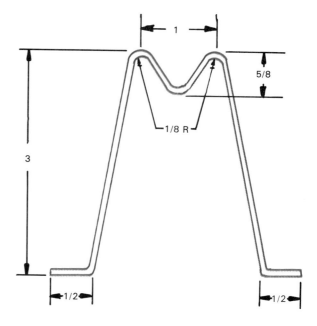

Fig. 4H. Motor bearing.

Chapter 5
Information and Technology

The information given in this chapter will help you to:
- ☐ List the three basic areas of knowledge.
- ☐ Define and give examples of scientific, technological, and humanistic knowledge.
- ☐ Discuss the use of symbols and signs as a way of communicating.
- ☐ Explain how signs and symbols are used by machines to pass and receive information.

Information is the same as knowledge. All that is known about any subject is sometimes called a **data base.** Building this knowledge base takes time. Something or someone must collect facts. For thousands of years people have been gathering and using information. They pass it on to every new generation.

This vast collection includes three basic areas of knowledge:
1. Scientific.
2. Technological.
3. Humanities.

Each of these areas is unique. Each is important in our lives. You will see this as we study them.

SCIENTIFIC KNOWLEDGE

Science deals with two areas of information. The first is what we know about the universe and everything in it. The second is the **theory** (what we guess is true) about the universe.

People who work in science are scientists. They search for a clearer understanding of our world. They want to know why things in nature act as they do. They also search for new facts. This takes many experiments and observations. The findings are then tested against old theories.

Scientific Method

Several basic methods are used to discover new scientific facts or to test older theories. Altogether, they are known as the **scientific method.** The scientist always begins with the idea that:
1. Everything that happens in nature can be understood. All you have to do is ask the right questions. Then you must do the right experiments.
2. Nature is always the same. Time or distance makes no difference. A scientist in California working on an experiment should get the same results as another scientist anywhere else at any time.
3. There is a relationship between a cause and its effect. Scientists design the experiment to produce a given effect. Then, they change the conditions one by one. This way they can find which conditions are producing the effect.

Scientists are always experimenting. There is no end to the kinds of experiments. On the one hand, they might launch a space probe to photograph a distant star. On the other hand, they might work in a laboratory studying molds or bacteria.

Experiments are designed to produce useful information, Fig. 5-1. Scientists seek answers to questions that are puzzling them. Experiments must often be repeated over and over to prove that the results are not accidental.

Developing a theory is another scientific method. A theory is an attempt to explain events that happen in our universe. A theory is not certain but the facts suggest that it is true.

Fig. 5-1. This scientist is experimenting with chemicals. (Rohm & Haas)

Developing theory is very hard to do. Often, the problem is lack of information. New facts give rise to new theories. Old theories may die because new facts disprove them. There are many examples from history. At one time, scientists did not understand burning. Experiments proved buring could not take place without oxygen. At another time, experts were puzzled that the theory of gravitation could not explain the movement of Saturn. Soon, astronomers found the planet Uranus which was affecting Saturn. See Fig. 5-2. Refer to Fig. 5-3 showing discoveries and applications.

	Diameter in miles	Mean distance from Sun in miles	Length of year (Time to complete one orbit)
Mercury	3,100	36,000,000	88 days
Venus	7,700	67,200,000	225 days
Earth	7,927	92,900,000	365 days
Mars	4,200	141,500,000	687 days
Jupiter	88,700	483,400,000	12 years
Saturn	75,100	886,000,000	29½ years
Uranus	29,200	1,782,000,000	84 years
Neptune	27,700	2,792,000,000	165 years
Pluto	8,700	3,664,000,000	248 years

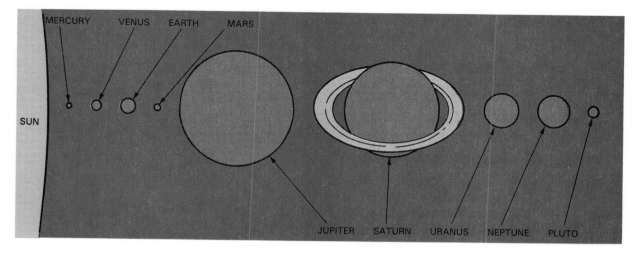

Fig. 5-2. Scientists produced this drawing of our universe. They were able to collect the information after they had observed and studied the planets.

MEASUREMENT

A very important scientific tool is measurement. Measurements create numbers which mathematicians can use. Sometimes theories can be stated mathematically to solve problems.

The whole electronics industry rests on scientists' theories about the action of electrons, Fig. 5-4. Throughout history scientific discoveries have been linked with technological advances. Technology puts nature to work. Science explains why technology works.

DISCOVERIES AND APPLICATIONS IN SCIENCE			
DATE	**DISCOVERY**	**BY**	**MODERN APPLICATION**
624-848 B.C.?	Relationship between sides of right angle triangle	Pythagoras	Truss designs in building
211 B.C.?	Area of a circle, principle of lever, screw	Archimedes	Machine design
Date unknown	Building, engineering	Romans	Domes, aqueducts, roads
Date unknown	Magnetic compass	Arabs and Chinese	Navigation
1446 A.D.	Movable type for printing	Gutenberg	Books & publications of all types
1536	Developed algebra and arithmetic	Cardan	Accounting, engineering, electronics
1583-1642	Principles of inertia, acceleration of falling bodies, statics	Galileo	Measuring force of gravity, astronomy, design of structures
1637	Analytic geometry	Descartes	Computation of rocket flight
1687	Law of gravitation Laws of motion	Newton	Space travel Jet motors
1705	Steam engine	Newcomen	Electric power generation
1774	Oxygen discovered	Priestly	Steel production
1798	Theory of heat as matter in motion	Rumford	Friction brakes
1799	Electric battery	Volta	Portable radio, electric starter
1831	Electromagnetic current	Faraday	Generators and motors
1843	Electrical, mechanical equivalents of heat	Prescott	Automobile motors
1887	Photoelectric effect	Hertz	Burglar alarm
1905	Relationship of space and time Mass-energy equivalence	Einstein	Studying nature of universe

Fig. 5-3. Since recorded time, humans have been gathering and using knowledge to make life better.

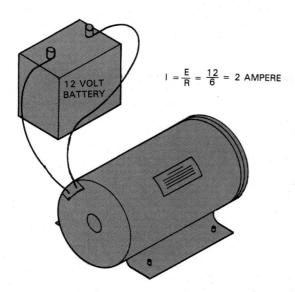

$$I = \frac{E}{R} = \frac{12}{6} = 2 \text{ AMPERE}$$

Fig. 5-4. A German scientist, George Ohm, studied how electricity acted. Then he worked it out as a mathematical formula. If the battery in the drawing delivers 12 volts of electricity and the resistance of the motor is 6 ohms, then the current equals 2 amperes.

BRANCHES OF SCIENCE

Usually the sciences are grouped around three basic branches:

1. **Physical sciences,** Fig. 5-5. These sciences deal with matter in its purest form. Physics, astronomy, and chemistry are physical sciences. Sometimes mathematics is included, but, generally, it is considered a common language or tool of all science.

2. **Earth sciences,** Fig. 5-6. These are concerned with the history, properties, composition, and behavior of our natural world. In this group are:
   a. Meteorology, the study of the atmosphere and all that happens in it.
   b. Oceanology or oceanography, the science of the seas.
   c. Geology, the study of the earth and, especially, its rocks.

3. **Life sciences,** Fig. 5-7. Life sciences study living matter. They look for the answer to many questions. How did life forms begin?

Fig. 5-5. A geophysicist, left, and technicians review section of a seismic map. They work for a major oil company.   (Getty Oil Co.)

Fig. 5-6. A geological consultant collects data in the rugged Rocky Mountains of western Wyoming. A support helicopter can be seen in the background.   (The Standard Oil Co. of Ohio)

Fig. 5-7. Life scientists study living matter. (Biomet Inc.)

How can living things take food and make it a part of themselves? How do plants and animals protect themselves? What is memory? This branch includes:

a. Botany, the study of plant life.
b. Zoology, the study of animal life.
c. Microbiology, the study of the smallest living things (algae, fungi, protozoa, bacteria, and viruses).
d. Medical sciences.
e. Agricultural sciences.

Technology

Technology is the knowledge of how to "do" or "make." It is knowing how to take information about our universe and use it to control the environment. An encyclopedia explains it as applying science to produce valuable products. The products are used to produce still other products or to make life better.

Without technology, agriculture could not feed the people of the world. Without technology food would spoil before it could be delivered to people. We also use it to locate minerals and petroleum. We use it to process these materials into usable products. We need technology to harness the atom and the electron for power.

Technology also produces efficiently. You have heard the phrase "time is money." It means that time is valuable. It should not be wasted. If a product costs too much to produce no one can afford it. Technology finds a way of "doing" faster and better. This is also called "efficient action."

Appropriate technology

In today's world, technology is also concerned with "appropriate" action. This means that the action must not waste resources (material). Appropriate action also means that there must be concern for the health and safety of people who make and use products. It means that there must be concern for the effect of the technology on the environment.

Technology has helped people control nature and build a better way of life. It has given us better food, better clothing, and better shelter.

It has made movement from one place to another easier.

Primitive people were not well equipped to deal with their world. They had little control over nature. To begin with, humans were small compared to many of the animals. Their tools were simple. They did not know how to raise animals and plants. Their lives were spent searching for food.

Early technology

Technology goes back into prehistoric times. First, humans adapted items in nature as tools. A rock or a branch became a tool. Then they may have learned how to control fire. This led to shaping of tools. With fire and water they could heat a rock and chip away the edges with cold water. This produced a cutting edge. See Fig. 5-8.

Fig. 5-8. Cave dwellers soon learned to use fire for such purposes as cooking food.
(American Assn. of Blacks in Energy)

Since their early concerns were for food, prehistoric farmers shaped implements to make agriculture easier. They tamed animals, Fig. 5-9, so they could farm more land.

By the Middle Ages, use of simple tools, fire, agriculture, domestication (taming) of animals, pottery, weaving, carts, ships, the pulley, and other mechanisms were well known. They were the basis for development of early science and more advanced technology.

Fig. 5-9. Since animals were stronger, humans trained them to help grow more food.

RELATIONSHIP OF SCIENCE AND TECHNOLOGY

Science and technology work hand in hand. One supports the other. While technology applies science, it cannot be said that science always precedes technology. In many instances the technology has come first and the explanation of why it works came at a later time. Early science came about because of humans' practical experiences at developing products. The explanation of the scientific principle came long after the crafts were developed. For example, the simple balance scale was used in Egypt for more than 1000 years before there was any knowledge of gravity. The Greeks had steam-powered toys long before there was any knowledge of physics.

Technology's Contributions

Throughout human history, technology has helped people in three important ways:
1. It increased their ability to produce large quantities of goods. Less time was needed to search for food. People had time to do other things. They developed skills in producing clothing, tools, and shelter. Trade developed as people exchanged their surplus (extra) products.
2. It reduced the time and labor needed to produce goods. Automatic machines process many materials. Early in the 18th century, most manufacturing was done by hand or on hand-operated machines. Workers were on the job 12 to 16 hours a day. Later, powered machines produced goods rapidly. The amount of human labor needed was sharply reduced. Today's factory workers usually work no more than 40 hours a week.
3. Technology has also made work easier. Coal mining and farming were much harder work before the use of machines. Farmers worked from dawn until dark. They could only produce enough food for about four people. Miners worked all day with pick and shovel to produce a few tons of coal. Moreover, the mines were dark, poorly ventilated, and dangerous. Today a farmer uses tractors and other labor-saving machines. One farmer produces enough food to feed 78 people.

The coal miner's work is also much easier and productive, Fig. 5-10. A mining machine operator can dig more than a short ton of coal every minute, Fig. 5-11.

Humanities

Humanities deals with human culture and backgrounds. Humanities wants to know about

Fig. 5-10. Coal mining is easier today because of better technology. The first class miner at right checks mine records as worker looks on. (Atlantic Richfield Co.)

Fig. 5-11. A continuous mining machine. Cutter head rips coal from seam. Large bucket (below) is called a gathering head. It scoops up coal loosened by the cutter head. (Lee-Norse Co.)

Fig. 5-12. A class in communications deals with written language. They discuss the news of the day from a local newspaper. (Chicago Tribune Co.)

Fig. 5-13. These women are archaeologists. They are working on an excavation in Alaska. (American Petroleum Institute)

our values, feelings, and ways of expressing emotion.

In our schools, the humanities are known as the "liberal arts." They open or free the mind to all of humankind's ways of expressing itself.

The humanities include many different areas of knowledge:

1. Languages. These are the sounds and signs people use to communicate. The sounds are speech. The signs are writings. See Fig. 5-12. There are about 3000 different spoken languages in the world.
2. Linguistics (a scientific study of languages). It describes language development and traces the history of language.
3. Literature (writings in prose or verse meant to entertain).
4. History (written record of important past events).
5. Jurisprudence (science or philosophy of law).
6. Philosophy (a study of the processes behind what we think and how we act).
7. Archaeology (the study of past civilizations through study of old ruins). See Fig. 5-13.
8. The fine arts. Painting, music, dance, sculpture and architecture.

PASSING ALONG INFORMATION

Knowledge has been accumulating (gathering) for thousands of years. At first, it was passed from generation to generation by word of mouth. It was memorized. Some people made it their life's work to remember the information. They acted as historians.

When writing developed, the information did not need to be memorized. It was stored in manuscripts and books. Knowledge has been expanding at a dizzying rate. It is doubling every five years.

However, we can store and pass along even more information through the growth of technology, Fig. 5-14. Machines now gather and flash information around the world in moments. We call such activity INFORMATION PROCESSING. That is, information can be considered a machine process.

Fig. 5-14. Machines help us process and pass along stored information.

INFORMATION AS A PRODUCT

Information can be thought of as a product. Machines can process facts like you might turn iron ore into steel.

The raw materials are the collected facts and figures. To convert this raw data into information requires processing. Processing involves collecting, recording, classifying, calculating, storing, and, finally, retrieving (getting out of storage). We call all of this DATA PROCESSING. You hear the term often today. Computers process the mountains of facts and figures we collect.

We store information in computer files, in books, on film, and in our minds. See Figs. 5-15 and 5-16. By itself, this information is of no value. We have to use it to satisy human needs or wants. Computers can help, but human intelligence is needed. It turns information into knowledge and knowledge into useful products or services.

Fig. 5-15. A layout artist works on pages that will be used in a textbook. Books are one way we store information.

Fig. 5-16. Information may also be stored and passed along with a computer. A computer operator uses a keyboard to store information in the computer. The monitor displays what is being stored. But the information is actually stored on a floppy disc. The disc has been placed in the disc drive so the information can be transferred. (Information for this book was stored in this computer.)

MACHINE STORAGE OF INFORMATION

It is easy to see how books can store information. Or how a movie can be stored on video tape and played back when we need it. However, computers store, compare, and recall information.

The Computer

A computer is a processing machine, Fig. 5-17. Using electronics, it processes information. It has four basic parts:

1. An input device. The device is usually a keyboard like the one on a typewriter. It takes the information and changes it into electrical signals the computer can understand. The keys act like switches. Electric signals can also be put on tapes and discs.

2. A processor. This is the brain that processes the information. It follows the program (set of instructions) fed into it.

3. A memory. This is a place where the information is stored. A memory will also store the program of instructions.

4. Output devices. These are parts that receive the messages from the computer. They do something useful with them. One type of output machine is a printer. It types out the message on paper. A television set or monitor is another output device often used. Other output devices are motors, loudspeakers, or other machines.

The parts of the computer are called components. They are also called **hardware** in computer language. The instructions for the computer are called programs and subroutines. Another name for all of the programs and subroutines is **software.** The software tells the computer what to do and how to do it. Without software the computer could not solve problems.

HOW COMPUTERS USE SIGNALS

How does a machine, such as the computer, process information? Let us take a simple example. A switch is a simple control that turns on a light. The switch completes an electrical path along which an electric current travels. When the current reaches the light bulb, it produces light. It will continue to produce the light until the current stops. You can make it stop by returning the switch to "off."

The flipping of a switch on and off could just as well be yes and no answers. You could also have numbers 0 and 1 mean "yes" and "no." The 0 could mean something is missing (off). The symbol "1" could indicate that something is present (on). These symbols are the numbers in a binary system. (Binary means "two numbers.") Are you beginning to see here a system of signals and symbols?

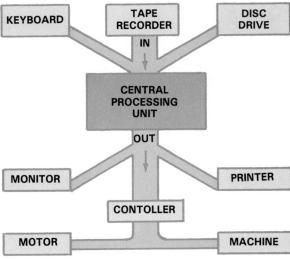

Fig. 5-17. Top. A computer receives, stores, and processes information. Bottom. A diagram of a complete computer system.

Think of how a light might be used as a signal. Suppose you wanted to arrange a signal at your house. Leaving a light on would mean one thing. Shutting it off would give an entirely different message. See Fig. 5-18.

You can think of the computer as a signal maker. In its memory are many tiny parts called transistors. These are small electronic parts that react when an electrical pulse reaches them. Some will reject the pulse. These are the "no" answers or zeros. But other transistors will receive the pulse. These are the "yes" responses or "1" signals.

With this simple system you could make a computer make decisions or give answers. This is how it would work:

Suppose your school was having a fair. To attend the fair, you've decided, a student must:
1. Live in the local area.
2. Pay an admission in advance.
First, you'd give the computer the names of all the eligible students. This would cause one of the transistors to accept the electric pulse whenever one of those names is again fed into the computer memory. Next, you'd enter the names of students who have paid the fee. This would cause a second transistor to accept the electric pulse whenever one of these names is fed into the computer memory.

Now the computer is ready to work. As students come to be admitted to the fair, their names are fed to the computer. If both transistors accept electricity, the computer will signal a "yes." If only one transistor accepts electricity, it will signal a "no."

This is a very simple explanation of how the computer works. There are thousands of such transistors in the computer making "yes" and "no" signals. It can make thousands of decisions almost instantly.

Understanding Signals

Many signs or signals are understood by our senses. Some are visible. A policeman motioning traffic through an intersection uses a visible signal. Signals can be sensed in other ways too. We can touch one another or shout to get attention.

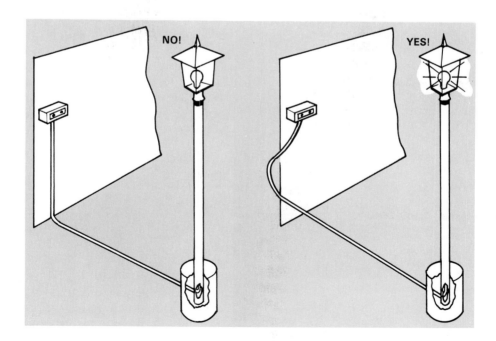

Fig. 5-18. Imagine how you could use a light to signal a "yes" or a "no" answer to a friend.

A machine tool in a factory can also receive signals, Fig. 5-19. It can be designed to respond to small pulses of electricity. When the signal is received, the machine will operate.

It may seem strange that a tiny current or a series of numbers can be an instruction to a machine. But stop and think how children learn to talk. They come to "sense" that certain sounds are related to objects, such as pets or people. It becomes fixed in their memory that one person is mother and another person is father. They are also able to give back that information by saying "daddy" and "mommy." The sound is not the person. It is a symbol which stands for that person.

It is just as easy to give a machine symbols which stand for the language symbols we have learned. Remember the switch we talked about earlier? A computer has many switches. They control current to feed certain arrangements of on-off signals into its memory. These signals stand for certain words and numbers. It works like Morse Code where arrangements of dots and dashes stand for letters of the alphabet. To the machine, the code, or sequence of on-off signals, has meaning.

Some information is transferred for communication. Other information is transferred to control some action. Both humans and machines respond to signals and symbols. Humans can use them to communicate with machines. Machines use them to communicate with each other.

CODED INFORMATION AND CONTROL

Machines can store information through coded messages sent to them. A music box is a simple example of information or signals stored on a drum or disc. As this device rotates, small steel projections (stored music) strike metal reeds causing them to vibrate and sound the proper musical notes stored on the disc or drum. It is a simple model of what, in computer language, is called ROM (read only memory). As the computer is "played" it sends the message to a device that can understand it. The device can be a machine tool such as a lathe. It can also be a printer. This is a kind of typewriter operated by a computer. See Fig. 5-20. It types out the message as a letter, a report, or a manuscript. The manuscript for this book was put on paper by a computer and a DOT MATRIX PRINTER. (This machine makes letters by placing dots on paper.)

Computers can be linked to other machines that respond by giving off radio or television signals. They can be linked to controls in an automobile that will feed fuel and air to the engine.

As we see, computers can be used to store and spread information. They can be used to control machines used in manufacturing. They can be used to do complicated math and make computations in seconds that would take a person hours to do. They can take new facts or data, compare them with previously stored information and provide new information.

Fig. 5-19. Two robots are spot welding an automobile part. A computer directs the robot arms to the correct point for welds. Another computer signal tells the robots when to weld. (Cybotech)

Fig. 5-20. A dot matrix printer places tiny dots on the paper in the shape of letters. It forms the shapes from directions it receives from a computer.

SUMMING UP

Information that is grasped by a person or persons is knowledge. All that is known about a subject is known as a data base.

There are three branches of knowledge. Scientific knowledge deals with our universe and how it works. Humanities is knowledge of human culture and backgrounds. Technology is applying scientific knowledge to change our environment.

Science deals with what we know to be true and what we guess is true about the workings of our universe. There are three branches of science: earth science, physical science, and life science.

Physical science deals with pure matter. Earth science deals with the history, properties, compositions, and behavior of our world. Life science is a study of plants and animal life.

Our ancestors had to learn about the universe little by little. They also had to apply that knowledge in a way that would improve their lives. They learned to pass along the information to other generations.

We can think of information as a product. The raw materials are facts and figures. We can collect facts and figures, process them, and produce new information. This is called data processing.

The human mind is capable of data processing. It can collect, sort, classify, and combine facts. It can create new information and then store it in the memory.

Today, machines have been developed to do all these things. In addition, machines can communicate with humans as well as with other machines. The machines that do this are known as computers.

KEY WORDS

These words were used in this chapter. Do you know their meaning?

Archaeology, Art, Binary system, Data processing, Data base, Dot matrix printer, Earth sciences, Hardware, Humanities, Information processing, Life sciences, Linguisitics, Literature, Philosophy, Physical sciences, Scientific method, Software, Theory.

TEST YOUR KNOWLEDGE
Chapter 5

Do not write in this book. Place answers to test questions on a separate sheet.

1. All of the knowledge about any one subject is known as a _____.
2. List the three basic areas of knowledge.
3. Describe each of these basic areas of knowledge.
4. A _____ is an attempt by scientists to explain events that we see happening in our universe.
5. Note the following areas of knowledge or studies and indicate whether they are related to science, technology, or humanities.
   a. Physics. _____
   b. Astronomy. _____
   c. History. _____
   d. Botany. _____
   e. Engineering. _____
   f. Literature. _____

6. Raw data processed into usable information by machines is known as _____ _____.
7. List the six activities included in processing information.
8. 0 and 1 are the only numbers in the _____ system.
9. Like humans, machines have signs that stand for objects. True or false?
10. A _____ _____ is a simple example of information or signals stored on a disc or drum.

1. Study the system of semaphore signals used by the U.S. Navy with a classmate. Use the signals to send a "message" to each other.
2. Use a minicomputer to solve a mathematical problem.
3. Study an abacus and then build a small counting device.

Machines are now important gatherers of information. An engineer uses a hand-held monitor as he checks out the electrical systems of a printing press. The monitor gives information instantly that would take precioius hours of inspection. (GE)

Machines became important to communication when the first printing presses were invented. Up to that time, information had to be handwritten. This press was used by Benjamin Franklin. (Smithsonian Institution)

APPLYING YOUR KNOWLEDGE

Introduction

Information is used to understand technology and solve technological problems. This information comes from many sources. It can be obtained by talking to people. Observing actions is another source of information. For most researchers, the printed word contains a great deal of information. This source may be (1) books, (2) magazines, (3) technical reports, (4) news releases, and (5) product catalogs.

This activity will let you gather and apply some technological information. You will be able to read about radio broadcast systems. Also, some diagrams and graphs will be pro-

vided. From these you will be asked to:
1. Outline the information.
2. Prepare a brief report.
3. Illustrate the report.

Procedure

1. Study the materials included with this activity.
2. Take notes on the important information.
3. Provide a brief outline for the information using the following headings:
   a. Historical background.
   b. How radio signals are broadcast and received.
   c. Importance of radio.
4. Write a brief report with illustrations (sketches) on radio broadcasting systems.

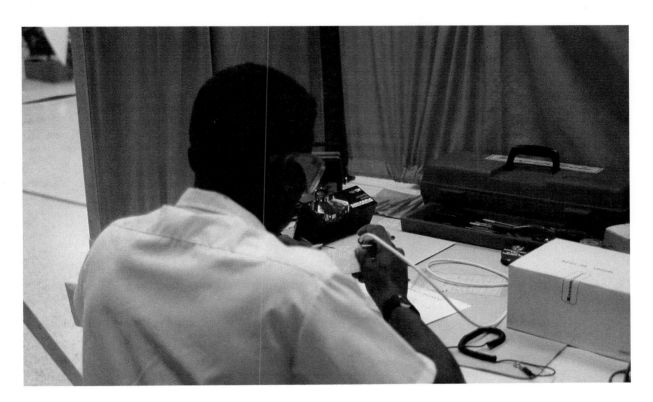

Experimenting with electricity and electronics can lead to a satisfying career.

RADIO'S EARLY YEARS
Robert A. Smithson

The radio is an invention of the 20th Century. In 1906 voices were first sent over the airways. In the United States, R.A. Fessenden built the first system. It used existing transmitters and receivers. This equipment had been developed for the wireless telegraph. But the telegraph sent only "dot and dash" signals. Now voices could be heard.

Still, radio, as we know it, had to wait. Better receivers were needed. These were based on the vacuum tube. This 1906 invention was developed by Lee De Forest.

The earliest home radio receivers were developed by Westinghouse Electric Company. Dr. Frank Conrad, a Westinghouse engineer, built a 200 watt transmitter in 1920. It was placed on the roof of a Westinghouse building in Pittsburgh. The transmitter became the first commercial radio station. It signed on as KDKA on November 2, 1920. Its audience was probably fewer than 1000 people.

HOW A RADIO WORKS
C. M. McCahall

Electric currents are electrons moving on a conductor. They were first used to send a message by the telegraph. This technological device was invented by Samuel F. B. Morse. Later, Alexander Graham Bell used them for his telephone. Both of these devices have a great disadvantage. They require wires to connect the transmitter (sending unit) to the receiver.

Electrons have a very important property. They can be caused to pulse back and forth very rapidly. Let's say, this is done in a coil at 30,000 pulses (cycles) per second. A coil some distance away will also pulse at 30,000 cycles per second. What happens is that the first coil sets up radio waves. These waves travel through the air. When they reach another coil they create the same pulse.

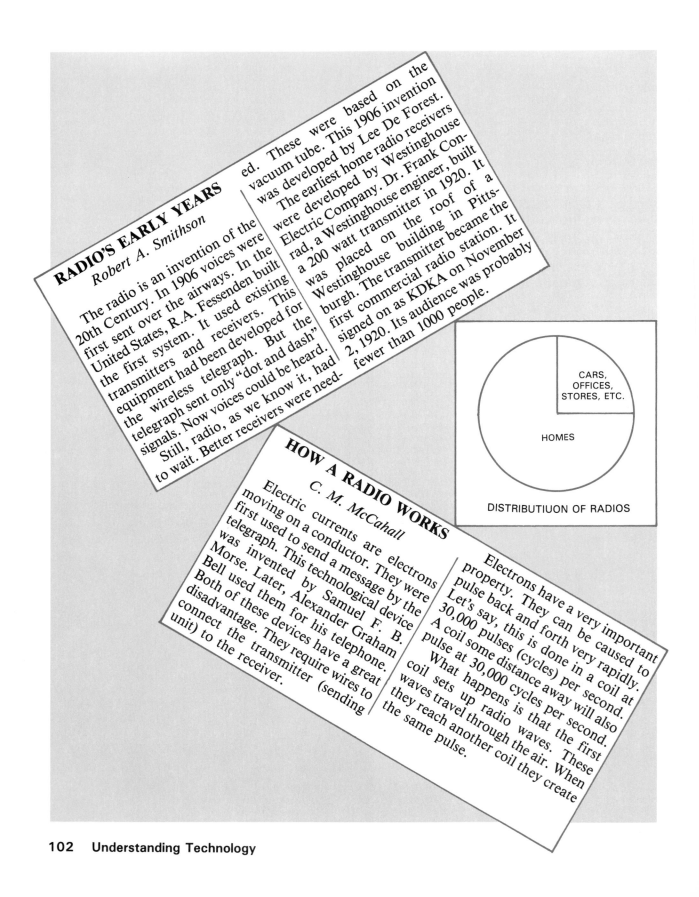

CARS, OFFICES, STORES, ETC.

HOMES

DISTRIBUTIUON OF RADIOS

RADIO TODAY

F.M. Wriack

Radio grew very rapidly during its early days. The 1930s and 1940s saw the medium become a major entertainment system. It had its own stars. Its personalities became household names. Big bands and individual musicians became famous. Plays, comedies, operas, sports broadcasts, and continuing stories were offered.

But after World War II, television became the dominant broadcast medium. Radio could have disappeared. But it did not. Instead it changed its emphasis. Out went the radio dramas and soap operas. Gone were the quiz shows and other entertainment shows. In their place came the disc jockey, frequent news and weather reports, and the call-in talk show. Radio adapted to a new way of life.

Today, there are nearly 9000 radio stations in the United States. About 1000 of these are noncommerical stations; they are not organized to make money. Many of these are part of the National Public Radio Network. Of the remaining stations, about 4500 are commercial AM stations and 3200 are commercial FM stations.

It has been estimated that there are about 450 million radio sets in America. About 75% of these are in homes and the other 25% are in cars, offices, stores, etc.

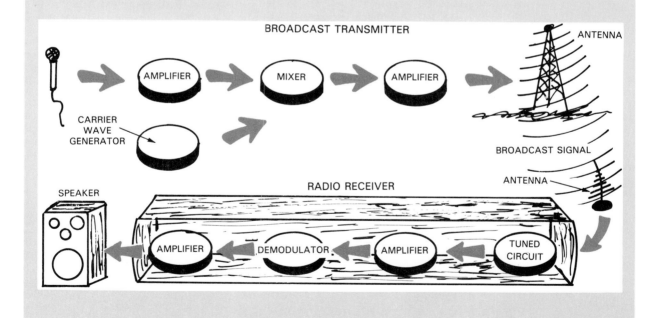

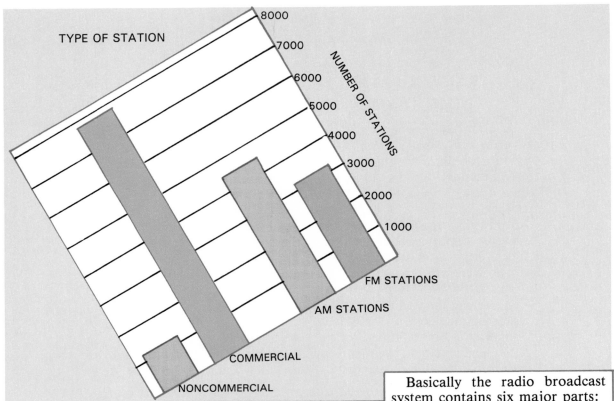

TYPE OF STATION

NUMBER OF STATIONS

8000
7000
6000
5000
4000
3000
2000
1000

FM STATIONS

AM STATIONS

COMMERCIAL

NONCOMMERCIAL

BROADCAST BANDS

A.M. Johnston

The wavelength of all radio stations is controlled. Wavelength is set by regulations of the International Telecommunications Union. These rules are designed to keep the signal from one station from interfering with that of another station. AM stations use frequencies between 535 and 1605 kHz (kilohertz or 1000 cycles per second).

FM stations use frequenices between 88 and 108 MHz (Megahertz or million cycles per second). Other frequency bands are used for two-way radios (police, taxi, citizen band, etc.), television, short-wave radios, radar, microwave transmission of telephone messages. messages.

Basically the radio broadcast system contains six major parts:
1. A microphone to change sound waves into electrical (electron) pulses.
2. A broadcast transmitter which combines the electrical pulses with a carrier (transmitter) wave. The transmitter then amplifies (increases the strength) of the signal.
3. A broadcast antenna (tower) which radiates (sends) the signals into the air.
4. A radio (local) antenna which receives the broadcast waves (signal).
5. A radio receiver which separates a selected radio wave from other waves. It then amplifies the signal and changes it into electron pulses.
6. A speaker which changes the electron pulses into sound waves.

Chapter 6
People and Technology

The information given in this chapter will help you to:
- ☐ Discuss some characteristics of individuals that would suit them for jobs in a technological world.
- ☐ Discuss how people move up to better jobs through work, education, and experience.
- ☐ Explain what skills are and how they may suit a person to a particular job.

We are in the middle of a great growth period. We have so much knowledge of how our universe was formed and works. We also are developing more skill at putting this knowledge to work through technology. With this growth we need more people with special skills and knowledge.

A CHANGING SOCIETY

We are also living in a rapidly changing society. Our country has already passed through several stages. Each of these stages saw great changes in what kind of work people did. During pioneer days and up to about 1900, we were an agricultural socity. Most people farmed, Fig. 6-1. This was our main industry. There were few careers a person could follow. It was called the agricultural age.

Later manufacturing became an important industry. This stage came to be known as the industrial age. Machines were invented to help make products and erect buildings. People were trained to run the machines and use construction equipment.

Fig. 6-1. At one time many people worked as farmers. It took more human labor to raise food. Most of the work was done by hand and with horses or mules. This farmer had to work long hours to get his work done.
(American Petroleum Institute)

People and Technology   105

After World War II it became more important for people to give information to one another. We had many inventions that helped us exchange information easily. The telephone became more and more important. So did radio. Books, magazines, and newspapers had become plentiful and cheap. Television was developed around 1930. It came into general use during the 1940s and the early 1950s. Now computers process information and make calculations for us, Fig. 6-2.

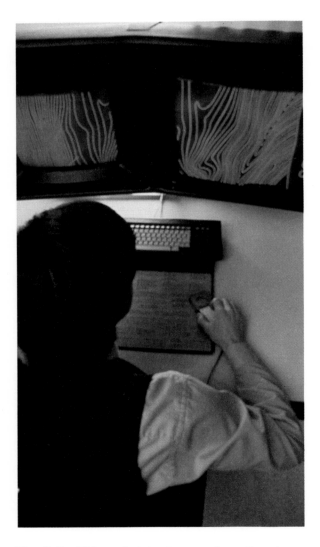

Fig. 6-2. This technician is part of an oil exploration team. She uses a computer to study likely spots for drilling.   (Standard Oil of California)

Experts are predicting even greater advances in knowledge between now and the year 2000. More new discoveries will be made than in all of human history. How will this affect you?

Each of us is a special person. We each have special abilities, Fig. 6-3. We sometimes call them talents. You may be very good in math. Or, you may excel in communicating skills such as languages or grammar. Your friend may do well in science courses.

GOALS, INTERESTS, TRAITS

Each of us will use or develop some type of technology. Where and how you and your friend will work will depend upon:
1. Personal goals. Not all of us want the same things out of life. Some want to live and work among their friends. For example, you may want to live in the community where you grew up. Perhaps, being home every night is important to you. On the other hand, you may prefer traveling, working with new ideas, and making new friends.
2. Interests. If you have held jobs, you may already know what you prefer. You may like working with others, Fig. 6-4. You get along well as part of a team. Then you should consider a job where you meet and work with people.

Or you may prefer working alone using tools, instruments, or a computer. This could indicate that you should look for a career working with tools or machines. A job where you will be alone most of the time may be attractive.

It is good to identify those courses you enjoy. They indicate where you will be most successful. A love for and ability to do math is important for accounting or engineering careers. On the other hand, communication careers depend on English and grammar abilities. An interest in several areas may lead to a career where these skills and interests can be combined. For example, an interest in writing and working with people or machines is important for authors, teachers, or editors. Such persons may work

Fig. 6-3. Each person has certain abilities. Each also has likes. The people pictured here developed their abilities into careers. A—Left, computer operators. Right, energy company employee works on off-shore drilling rig. B—Vehicle operator with the front end loader she operates. C—Drafting is preferred by this man. D—This woman works as a coal miner. (American Assn. of Blacks in Energy; Gulf Oil Corp.; Conoco Inc.; Atlantic Richfield Co.)

Fig. 6-4. If you enjoy people you probably should consider a job where you work with others. This woman is a supervisor for a gas company. She directs the work of others. (American Assn. of Blacks in Energy)

in areas like auto mechanics, carpentry, or printing. See Fig. 6-5.

3. Physical traits. Careers make demands on one's mental abilities. But they also call on our physical abilities. Certain jobs may require long hours of physical activity. Others will require above average strength. Still others need nimble fingers or quick reaction time. Others may depend on good eyesight.

Regardless of your abilities, or **physical characteristics** (those things that make us good at certain tasks or endeavors), you should take pride in them. Never hide them, but try to improve yourself through work and practice. They may be the door to your future work.

GETTING CAREER INFORMATION

There are many ways you can learn about jobs that interest you. The world of technology offers many satisfying and challenging jobs. One of the best ways is to look for books on careers in your school library, guidance office, or the public library. They will describe jobs in any number of fields. They also tell you what is needed to get and hold the job of your choice.

Talk to people doing the kinds of work you think you would enjoy. They can tell you what the job is like. Most will be flattered that you show an interest in what they do.

Chapter 5 talked about the three areas of knowledge: science, technology, and humanities. You learned that in technology, people apply know-how to a job. There are many jobs to be had in this field. The areas include: engineering, teaching, journalism, editing, management, supervision, and skilled trades of every description. These occupations provide great satisfaction to those who work in them. The skills required are as different as the job names themselves.

It is customary to divide jobs into categories that have something to do with the kinds of skills required. These categories include:

1. Administrative, managerial, and professional occupations.
2. Clerical occupations.
3. Service occupations.
4. Agriculture, forestry, and conservation.
5. Bench work.
6. Structural work.

Fig. 6-5. Teachers are persons who like to work with others. Often, they have interests in many different fields.   (Atlantic Richfield Co.)

Teamwork

In most industries teamwork is important. Each member of the team has a special job to do. Still, everyone must work together toward the same goal.

It is somewhat like a basketball team. The club owner and his or her staff provide leadership. They provde the financing and set policy for the coach and players. But they do not tell the team how to play the game. The coach plans plays to reach the goal: winning the game. Fig. 6-6 shows another kind of team doing its job.

made. Then the technologists bring in the technicians and skilled workers. They work out the problems. They will build the body.

The technicians explain the idea to the drafters and other skilled people on the team. The drafters prepare working drawings. Other workers make the parts. Still others assemble a prototype (model). Then the technicians will test the prototype, Fig. 6-7. Often parts must be redesigned. New materials may need to be found. Changes will be made. Work continues until the design functions as the engineer intended. It can then be put into production.

Fig. 6-6. These students are a team too. They are in a planning session for energy conservation activities. (Atlantic Richfield Co.)

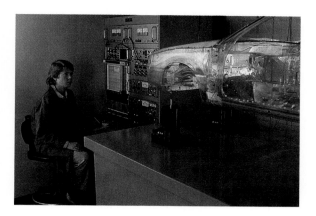

Fig. 6-7. Technicians are part of the team that produces products. They are always involved in the building and the testing. This woman is testing a plastic prototype of a new automobile body. (Chrysler Corp.)

The engineer is the coach of the technology team. The technologists are like the assistant coaches who supervise and train the team members. The technicians and the skilled workers are like the players on the basketball team. They carry out the special jobs that make a team a winner.

Suppose the job at hand is to design a new plastic automobile body design. The engineer gathers information and develops ideas. He or she draws sketches. Following good engineering principles, the technologists work with the engineer. They determine shapes, strengths of materials, and how finished parts are to be

Team Effort

It is important to remember that every team member supported the effort to build the new auto body. Had any member not done his or her job the team effort would have suffered. Perhaps the project would have failed. Team effort is as important in the world of work as it is in sports.

Making our technological system work takes many different kinds of abilities. You can think of a company as a system of people of varying skills. Each needs the others to make the system work. All are important.

Activities that go into making a product, providing a service, or building a structure are called **enterprise.** An enterprise is also another name for a company.

Starting a company takes planning, work, and organization, Fig. 6-8. It means bringing together all the resources needed to make the product. This includes getting the financing (money), a building, machines, workers, and materials.

Leadership

Leadership is the ability to lead or guide other people. It means other things too. A leader knows what needs to be done. But, more than that, he or she can organize activities. In addition such a person is able to get other people to take directions and follow them. Such a person has the confidence of others. They believe that the leader can bring the activity to a successful end.

Remember the example of the basketball coach given earlier? The coach, along with his or her assistants, provides a special kind of leadership. They get the team to work hard toward one common goal.

MANAGEMENT

All enterprises need management. These are people who organize the resources to make the product quickly and efficiently. They want a product to be built well. It must also be affordable.

Managers

Managers are people with special leadership qualities. You may know people who get along well with everybody. They are good at getting others to cooperate on a project. They are good at organizing people to get a job done. They are not afraid to make decisions. Usually their decisions are good ones. They have great determination and confidence even when things are not going well. They seem to enjoy the challenge of difficult jobs. These people have the traits of good managers and leaders.

The top managers in a company are the officers: **presidents, vice presidents,** and **directors.**

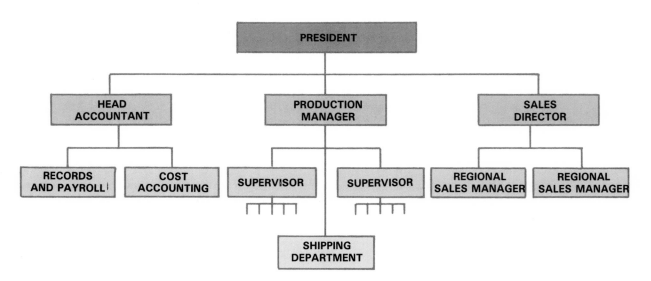

Fig. 6-8. A small company might be organized something like this. The lines coming from the president show that the head accountant, the manager of production, and the sales director report to (work for) the president.

They are the ones who set the company goals, Fig. 6-9. They direct the operation of its departments. They decide the directions the company will go.

Working under them are **middle managers.** These people direct most company operations. Middle managers oversee production, marketing, finance, personnel, and engineering activities.

The lowest level of management is called operative or first-line management. These managers help plan the day-to-day operations of the business. They also set up the work schedules and supervise production workers, salespersons, bookkeepers, and other employees.

The Scientists

Science is a separate field from technology, Fig. 6-10. The scientist's work is to find out why the natural world works as it does. Scientists discovered how to split the atom to release great quantities of energy. However, they did not find ways to use this energy. This is the challenge to technology.

Young people likely to be successful as scientists enjoy physics and earth science courses. They also tend to like mathematics and chemistry. They feel at home working with delicate laboratory instruments and performing experiments. They also tend to like the exactness and detail that goes with such work.

Fig. 6-9. Top. A company manager (center) conducts a staff meeting. (American Assn. of Blacks in Energy) Bottom. VICA (Vocational Industrial Clubs of America) stresses leadership training for students. This young woman was a winner in a national skill "olympics" speech contest. (Jeniene Ratz)

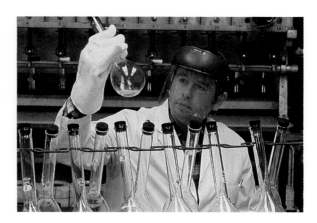

Fig. 6-10. Scientists experiment to learn why the universe and the things in it act as they do. (Anheuser Busch)

Engineers and Technologists

Engineers and **technologists** are the ones who use the knowledge of physical and mechanical principles and mathematics. They design processes, products, and structures that make life better. They use their knowledge to:
1. Produce power.
2. Build roads, bridges, and skyscrapers.
3. Design machines.
4. Set up manufacturing systems.
5. Change materials to make them more useful.

The engineer is the leader of the team. She or he often directs the work of the technologist. This person may be a specialist in product or process design. The technologist is a fairly new type of professional person. As engineers became busier in design, technologists took over some of their duties. They work between an engineer and a technician.

Technicians

Some people like to know how machines work. They are always tinkering with machines. These are the traits of a good **technician.** Such a person may not be interested in creative design. They solve problems at the operational level. Generally speaking, the engineer knows the theory of how and why something works. A technician knows how to make it work.

Technicians work in many different fields, Fig. 6-11. The greatest number work in the

Fig. 6-11. Technicians work in many different fields. (Clorox Co.)

health field. They operate and repair delicate medical instruments and devices. There are about 20 different kinds of technicians who work in the health field.

Over 60 different kinds of technicians work in the fields of engineering and other technologies. Some work on computers, robots, and automatic machines.

Skilled Workers

Working with professional engineers, technologists and technicians are many skilled workers. These people construct and assemble various products. They are the people who actually manufacture or build what others have designed. They may work on assembly lines designed by engineers and technicians. They perform the skilled operations required to build a road, a dam, or a skyscraper. They operate the machines to make parts. They prepare the patterns for casting materials. They operate road-building machines, cranes, and build construction forms. They must do their jobs safely, accurately, and efficiently. The finished part or structure must be safe, durable, and useful.

Generally speaking, the crafts can be organized into several areas:
1. Foundry occupations. These include patternmakers, molders, and coremakers.
2. Machining occupations. Machinists, instrument makers, machine tool operators, set-up workers, and tool and die makers make up this group.
3. Printing occupations. Skilled people in this group include bookbinders, compositors, lithographers, photoengravers, and printing press operators.
4. Other industrial production and related occupations. An example of the workers in this field include assemblers, painters, production supervisors, boilermakers, electroplaters, forge operators, millwrights, and welders. See Fig. 6-12.

Choosing and Advancing in a Career

Deciding on a career is important. It is not like any other activity you do in school. Because it is so important, you will need to take your

A   B

Fig. 6-12. Left. An engraver prepares a plate for a greeting card. (American Greeting) Right. A worker assembles photovoltaic cells at a research center. (Standard Oil Co., Indiana)

time. Try to learn as much as you can about each kind of work. Match the requirements and major responsibilities with your own abilities, interests, and values. Also pick out the requirements that do not match your characteristics. Decide if there is something you can do to remove the mismatch. If not, then the career may not be for you.

As you get more information you will be able to eliminate some careers. This will leave you with fewer to investigate. Do not be in a hurry as you see others making choices. Some make career choices very early. Others are adults before they decide.

It is also important to remember that a choice does not have to be final. You can rethink a decision and change it.

Changes in technology are opening up new jobs every day. At the same time, old occupations are being eliminated. Persons entering the job field today need to be flexible. This means they must be willing to change jobs. Many people will make career changes five times during their work life.

The Career Ladder

The career ladder, Fig. 6-13, means working your way up to better jobs as your skills and knowledge improve. Usually people move on to related jobs. Each job gives added responsibility and more pay. Sometimes the job promotion and training come from within the company. Sometimes it is an offer from another company. As you decide on a career,

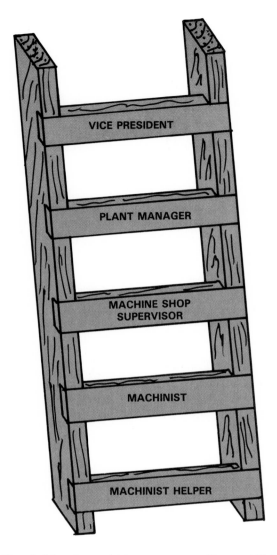

Fig. 6-13. An ambitious person will use abilities to get ahead. This is called climbing the career ladder.

Fig. 6-14. Improving skills is an important part of advancing on the job. Many companies provide valuable training for people who have the desire and the ability to advance.　(GM)

you should think ahead to the chance for promotion. See Fig. 6-14.

It is important to show your superiors that you have developed new abilities. Also, you must show interest in taking on more responsibilities. Being on time for work, doing the job with enthusiasm, and responding to job challenges are important. They show your supervisor that you are mature and trustworthy.

It is wise to choose a job that gives you a chance to move to a better job. Many beginning jobs are the first step to a better one. A successful assistant editor can move up to editor-in-chief. A good accountant may become the manager of the accounting department. An ambitious engineer may become general manager of the company. Some persons start in a company with a low-paying job and rise to become the company president. Others may start their own businesses. They become entrepreneurs.

Entrepreneurship

Entrepreneurs are an important part of everyone's life. There are many of them in every community. They are people who:
1. Plan and organize a company.
2. Direct its operation.
3. Shoulder the risk of ownership.

A business is formed to provide either a product or a service. For example, a brick factory manufactures a product—bricks. Brick masons provide a service. They lay brick to make walls or other structures. See Fig. 6-15.

Think of all the small business people in your neighborhood. They provide important benefits to the community. Not only are they easy to find when we need goods and services, they also provide jobs. About 58 percent of the jobs in the United States and Canada are in small businesses. Some 440,000 new small businesses are started every year.

Organizing a business

Businesses must be carefully planned. Even before they are opened there is a great deal of work to do. The owner must decide what kind of product or service to offer. Questions must be answered. Problems must be solved. Should I buy or rent a building? Where should the business be located? There must be a plan for purchase of materials and machines. Someone must hire and train workers. Work procedures and record keeping must be organized. Methods must be worked out for reaching and attracting customers. The owner must handle hundreds of such details.

Fig. 6-15. A self-employed brick mason provides a service. A building contractor will pay him to construct walls.

Directing the business

Directing a business means taking charge of its day-to-day operation. Entreprenuers set up their own work schedules. They also direct the work of others. They make the important decisions about what must be done and how it is to be done.

Shouldering the risk

Starting a business is always a risk. About 27 out of every 100 fail within the first three years. Entrepreneurs accept the responsibility for failure. If they make a bad decision, they take the blame. They promise to pay off the debts if the business fails. They must replace bad products when customers complain. They guarantee good service. They take the responsibility and make good on employees' mistakes.

Qualities of an entrepreneur

Being a successful entrepreneur takes special qualities. Good health is important though not always essential. Running a business usually means working long hours. Often, too, heavy physical labor is necessary.

Another mark of good entrepreneurs is that they are "self starters." This means they do not need others to direct their work. They can see what needs to be done and they do it. They are not afraid to make decisions. They have great confidence in their ability to succeed.

True entrepreneurs are willing to take risks. They will leave the security of a good job to be on their own.

While they will risk failure, they leave little to chance. They plan carefully for the success of their businesses.

SUMMING UP

We are in a period of history when knowledge is expanding at a high rate. In the next 15 to 20 years knowledge will advance more than it has in all of recorded human history.

How you will fit into this changing world depends upon personal goals, interests, abilities, and physical traits. Finding the right kind of work will require reading about careers and asking questions of people in different occupations. You will have many fields from which to choose. Some jobs will require skills of managing people; others will demand creative abilities. Still others will require skills of making parts and handling tools.

Whatever career is best for you, you should consider the opportunities for advancing to more responsible work. This is known as "moving up the career ladder."

Some people prefer to start their own businesses. They have confidence in their own abilities. Moreoever, they are willing to take risks for the rewards offered by entrepreneurship.

KEY WORDS

These words are used in this chapter. Do you know their meaning?

Age of information, Career ladder, Directors, Enterprise, Entrepreneurship, Foundry occupations, Machining occupations, Middle managers, Physical characteristics, President, Technician, Vice president.

TEST YOUR KNOWLEDGE
Chapter 6

Do not write in this text. Place answers to test questions on a separate sheet.

1. Explain why we are in a period of great growth.
2. Our country has passed through several stages in its development. At one time most people were engaged in farming. This was known as the _____ Age. Other stages in our development are known as the _____ Age and the _____ Age.
3. Name three things about you that will help you decide where and how you will work.
4. Why is teamwork important when you are working for a company?
5. People who organize the resources of the company to make the product as quickly and inexpensively as possible are the _____.

6. A(n) _____ knows why something works and can design it; a _____ knows how to make it work.
7. What do skilled craftpersons do?
8. What is the career ladder?
9. A(n) _____ is someone who will start and operate his/her own business.

SUGGESTED ACTIVITIES

1. Draw up a chart listing your abilities, interests, and values. Include what you are good at doing, as well as what you dislike. Try to be honest and look at yourself as you think others might see you.
2. Find several careers you think you might like. Gather information about the requirements of these careers. Describe the duties of each.
3. Compare your abilities, likes, and strengths with the career information you have collected. Do not expect a perfect match; however, the effort will start you on the way to career planning.
4. Invite a local businessman to speak to your class about his experiences in starting a business.

Tasks you like to perform are a good indication of what you would like to work at for a career. (Coachman Industries)

APPLYING YOUR KNOWLEDGE

Introduction

Each person has unique skills and abilities. Some people can think through complex problems. Others can build beautiful objects. Still others have great art ability. Some individuals can organize and manage people. Others can easily operate equipment and machines.

Jobs, likewise, are different. They require a mix of human skills. These skills can be divided into six major groups:

People skills. Ability to get along with people. Requires an understanding of human nature. People with high people skills can accept others with different abilities and attitudes.

Mental skills. Ability to analyze and solve problem using thought processes. Mental skills require a person to think clearly. Such a person must be able to use the mind to analyze complex situations.

Physical skills. Ability to use the human body (hands, arms, etc.) to complete a job. Usually physical skills require physical strength. Also, they usually require people to have good hand-eye coordination.

Management skills. Ability to organize and manage tasks. Management skills require the ability to separate complex jobs into smaller tasks. Management skills include a special people skill—the ability to get people to accept and work toward a specific goal.

Artistic skills. Ability to use the mind and the hands to create pleasing objects. The ability requires the use of design principles (color, balance, harmony, etc.) in creating an object.

Communication skills. Ability to convey thoughts and ideas to others. This ability requires skill in using words, symbols, and drawings to convey messages.

Each of us must develop an understanding of our abilities. Then we need to know what jobs can best use these abilities. This activity will allow you to look at pictures of people doing a variety of jobs. Then you will work in groups of 2-4 students. You will be asked to rate each job in terms of the six types of skills listed above.

Prepare or use a form like the one below for each job. You will decide if the job requires high, average, or low levels of each skills. (Fill out a form for each of the 12 jobs shown in the pictures.)

JOB RATING FORM

Group:_____   Date: _____

Job Name: _____

Rate each job's requirements for the types of skills. Rate them as "high," "average," or "low."

Job Title: _____

People skills _____   Mental skills _____

Physical skills _____   Management skills _____

Artistic skills _____   Communication skills _____

Fig. 6A. Welder. (Westinghouse Electric Co.)

Fig. 6D. Electronic component assembler. (Avnet, Inc.)

Fig. 6B. Sales training officer. (Clorox Co.)

Fig. 6E. Tree faller. (Weyerhaeuser Co.)

Fig. 6C. Product designer. (Ohio Art Co.)

Fig. 6F. Company president. (Ohio Art Co.)

Fig. 6G. Production supervisor. (Ohio Art Co.)

Fig. 6J. Automobile assembler. (General Motors Corp.)

Fig. 6H. TV camera operator. (American Petroleum Institute)

Fig. 6K. Computer-aided drafter. (General Motors Corp.)

Fig. 6I. Heavy equipment operator. (John Deere Co.)

Fig. 6L. Electronic circuit designer. (Arvin Co.)

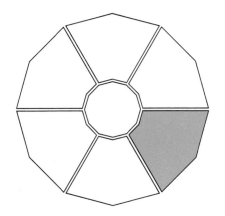

Technological Systems

If you have ever watched ants on an anthill, you may have wondered what purpose there was to their movements. Ants scurry here and there. Some are coming; others are going. Are they following some direction or design? We wonder.

You may well have the same questions about the activities of people around you. What is the purpose in their movements and activities? Is it organized in such a way that all of the things people work at are part of a larger picture? Or do we produce products and provide services with no master plan? Do we simply hope they will be useful to someone sometime?

Another question: what, if anything, do these working people have to do with one another? What does your neighbor who works with computers have to do with the person working as a carpenter? As a lathe operator? Or with the person who drives a city bus? Do these activities which seem to be so unrelated really tie into each other some way?

Actually, each of these activities is a part of a technological system. These systems are in your community. All are interrelated some way.

In Chapter 1 you learned what a system is. Now you are about to study several different systems. Each one is a complete system. Each is distinct but dependent on the others. Some of the same resources will be used but the outputs will be different.

It is the purpose of technological systems to change certain resources into end products. These products can be new, useful products, new and vital information, or energy to drive

technological systems. Or, the end product of a system may be to provide a useful service for people.

Every system includes inputs, processes, and output. Inputs are the resources used such as people, information, materials, and energy. The processes are the activities which change the inputs into something of greater value and use to people. The outputs are the items, services, and structures that make our lives better than before.

The following chapters deal with the various systems and their processes. These processes can be thought of as activities which transform the inputs or resources:

- Manufacturing turns raw material into automobiles, clothing, toothpaste, and thousands of items we use everyday.
- Construction transforms manufactured products and materials into structures we need for living and carrying on all of life's other activities. It produces houses, highways, bridges, tunnels, dams, and pipelines.
- Communication transforms raw data (information) into organized designs, meaningful instructions, and purposeful skills.
- Transportation transforms the energy in chemicals, the sun, and wind into power. The power moves people and materials so that all can share in the efforts of technology. It means that automobiles built in Windsor, Ontario or Detroit, Michigan can be driven by people from Regina, Saskatchewan or Lubbock, Texas. Or that food grown in Iowa can be moved anywhere in the world.

Design work is a form of communication. It processes information and tells others how a product is to be made. (Ohio Art Co.)

Construction uses manufactured materials (lumber, steel, and concrete) and produces a structure of them.

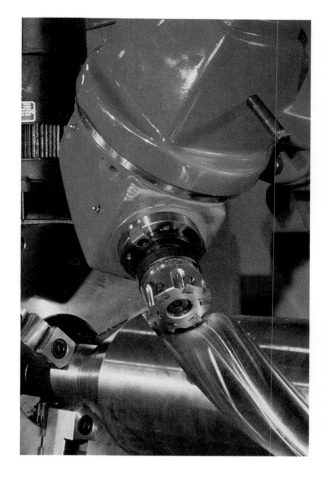

Manufacturing changes materials into more usable forms. (Kearney & Trecker Corp.)

Transportation moves people and materials from one place to another.

Data are often transmitted over telephones and computers.

Chapter 7
Designing Technological Systems

The information given in this chapter will help you to:
- ☐ Define design.
- ☐ List and describe the steps in the design process.
- ☐ Recognize and define a design problem.
- ☐ Gather information needed to solve a design problem.
- ☐ Generate solutions for design problems through rough and refined sketches.
- ☐ Describe the types of pictorial sketches.
- ☐ Use brainstorming to improve product designs.
- ☐ Refine designs through detailed sketches and models.
- ☐ Describe ways to test design solutions.
- ☐ Describe and prepare detail and assembly drawings.
- ☐ Describe and prepare bills of materials.
- ☐ Present a design for approval.

We are all affected by technological systems, Fig. 7-1. We use communication systems to send and receive information. We live in houses or apartments which are products of construction systems. We travel on transportation systems. And we daily use products from manufacturing systems.

However, these systems were not always here. Nor will they remain the same during our lifetime. Each technological system was designed. It was invented, developed, and refined by people. Also, the systems are forever being redesigned to make them work better. They are improved as technology advances, Fig. 7-2.

Each technological system is made up of a series of devices or products. For example, a transportation system consists of vehicles, guideways (roads, tracks, etc.), signal devices (signs, signals, etc.) and stations or terminals,

Fig. 7-1. Technological systems, like this satellite communication system, make our lives better. (Harris Corp.)

Fig. 7-2. Computer systems let us do work more quickly and accurately. (Clorox Co.)

Fig. 7-3. Each of these devices can be further divided into elements. A vehicle has a power source, a power transmission unit, a steering or control mechanism, support or structure components, and a guidance system.

All technological systems are made up of a number of parts. Each part must be designed and produced. This chapter will focus on designing parts and products for technological systems. The systems themselves will be explored in the chapters that follow.

Fig. 7-3. We all benefit from modern transportation systems. (CSX Corp.)

THE DESIGN PROCESS

A **design** is the plans for a device which solves a problem. The process is called engineering design if a physical item is being developed. Engineering design is used to plan buildings, products, roads, waterways, machines, and many other devices. As you study the items illustrated in this text, note that each one of them has been developed using the design process.

The act of designing involves a series of related steps. Each designer follows these steps as she or he completes a design project. These steps, often called the design process, include;
1. Defining the problem.
2. Gathering information.
3. Developing possible solutions.
4. Selecting promising solutions.
5. Refining selected solutions.
6. Testing design solutions.
7. Selecting best solution.
8. Interpreting the solution.
9. Presenting solutions for approval.

Each of these steps must be completed in the order listed. However, the designer may have to do some backtracking. He or she may have to go back to an earlier step to improve the process.

For example, a designer may be trying to develop possible solutions to a design problem. However, incomplete information may have been gathered at an earlier step. The designer may be having trouble developing good solutions. Therefore, the designer must stop and do more research. The missing information is gathered.

In another case, the problem may be redefined as more information is gathered. This leads to new solutions. A clearer picture of the design problem may develop as later steps are completed.

The process is like a flight of stairs, Fig. 7-4. The designer can climb from one step to the next. But if one step is found to be incomplete the designer can turn back. An earlier step can be improved or enlarged. Use of information from a later step to improve an earlier step is called **feedback.** Feedback loops improve the operation of any system.

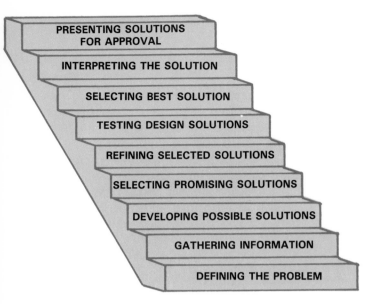

Fig. 7-4. Steps in the design process follow this order. Work from bottom to top.

Defining the Problem

All products and technological systems are designed to meet a human need or want, Fig. 7-5. Technology is appropriate only when it serves people. The entire design process should be directed to the primary goal: developing products and systems to make life easier.

The first step in the design process is to clearly define this need or want. This definition drives all other design activities. It gives all the other design steps proper direction.

The problem definition should include a clear statement of the task at hand. For example, the problem may be to "design a device to hold books upright." Note that the statement did not give the answer. It did not say: "Design a set of book ends." Book ends are only one solution to the first design statement.

Likewise, a statement to design a chair is too restricting. The designer would have to start with the definition of a chair: "a device with a seat, back, and four legs which holds a person in a seated position." A better statement uses the last part of the chair definition. It might read, "Design a device to hold a person in a seated position." The solution to this design problem *could* be a chair. But it could also be a bean bag or a wicker "basket" hanging in a frame.

To be useful, the definition must also include other information:
1. Describe who will use the product or system.
2. Tell where it will be used.
3. Give size and material restrictions (things you can't do) that may exist.
4. Suggest styling themes to be considered.

Therefore, a more complete product definition for the bookholder would be:

"Design a device to hold four to ten textbooks upright. The device should fit on top of a desk. It should blend with both Early American and Contemporary styles. The intended buyers are junior or senior high school students."

Or, a product definition for a toy might read like this:

"Design a set of wooden toy trucks. These toys should be appropriate for children four to seven years old. Each toy should be able to be used either inside or outside of a home."

Do these definitions start to give you an idea of what is wanted? What statements would you add to make them more complete?

Fig. 7-5. Products and technological systems are designed to meet human needs. (Ohio Art Co.)

Gathering Information

Designs are based on knowledge. The designer must have or must gather information before designs are developed, Fig. 7-6.

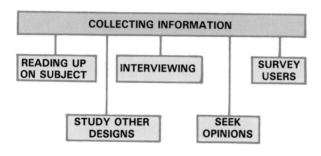

Fig. 7-6. Designs are based on knowledge.

Suppose you were asked to design a pencil holder. Could you start sketching possible designs? Do you know if it will be a wall or desk holder? What size are standard pencils? What color of holder do people want? How many pencils should the holder hold? Do people prefer wood, metal, or plastic? Will the holder be used in a home or office? You need such information to be a good designer.

Look at the bookholder design problem statement again. Do you need answers to the following questions to guide your design effort?
1. What size are textbooks?
2. What is Contemporary design?
3. What is Early American design?
4. What decorations would junior and senior high school students like on a bookholder?
5. Do the intended buyers like certain material better than others? Would they prefer wood, metal, or plastic?

Are there other questions that need to be answered? Would more information make the designing task easier?

How do designers get answers to their product design questions? There are several sources of information. These, as shown in Fig. 7-7, include:
1. Printed materials.
2. Customers.
3. Company records.

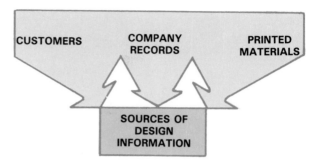

Fig. 7-7. Information for developing designs can come from many sources.

Printed materials

Books, magazines, and catalogs contain a wealth of information. For the designer, these sources can provide ideas for creating new, exciting designs. They can also show her or him how other designers solved similar problems.

Gathering information from printed materials can save time. Fewer poor designs will be developed. The ones that are developed are more likely to meet people's needs and wants.

Let's go back to the bookholder design problem again as an example. Using books and catalogs, the designer could study Early American and Contemporary design. Good examples of each could be photocopied for later reference.

Catalogs can also show how other designers have solved a similar problem, Fig. 7-8. A file of book holder designs can help the designer develop new and different solutions.

The designer may need to study desk designs too. The holder is to fit on a desk. It cannot take up too much room. One so large that a person cannot use the desk is worthless.

Customers

All technological devices and systems are designed for people. People must like them well enough to buy them. Therefore, customers opinions must be considered.

Often designers will ask customers for their opinion. The public may be asked what they want to see in a new device. Or they may be asked to try out a new product. Public opinion on color or size may be sought.

Fig. 7-8. Designers gather information from many media including books, magazines, films, and videotapes. (AMP, Inc.)

Fig. 7-9. Market researchers gather reactions of potential customers to a new product.

Different decorations may be developed. Then customers react to them. They may try out sample products made up from different materials. Selected people judge each one.

All these activities are called *market research*, Fig. 7-9. Research uncovers facts about the market: Who will buy a product? What features are wanted? What size and color is best for the product?

Early consideration of customers' opinions will make the design process more efficient. People will buy and use the devices.

Company records

Every company keeps records. Certain records are a great help to the designer. These records include customer complaints. They tell the designer what the customer does not like about existing products. A good designer will avoid these features in new designs.

Often the company keeps a file of letters asking if a product is available. These letters are good indications of products that are needed and wanted. Careful review of complaints and requests for products can trigger many good product ideas.

Developing Possible Solutions

At this point the designer knows what the design problem is. It has been carefully defined with the help of the information gathered. Now the designer is ready to develop possible solutions. (A final solution will come later.) The goal now is to develop several promising solutions. This approach improves the chances of having a winning design.

Designers generally start by sketching their ideas. They take what is in their "mind's eye" and put it on paper. This process has at least two major steps:
1. Developing **rough sketches.**
2. Developing **refined sketches.**

These steps start to move an idea along a path from a person's mind to a visual presentation.

Rough sketches

All people have ideas running through their mind. During the design process people try to think of many ways to solve a design problem. Ideas flash into the mind and may be lost very quickly. Good designers try to capture as many of these ideas as possible. They let the hand capture what their mind generates, Fig. 7-10.

The first sketches are no more than outlines and shapes. The designer will try to think of many ways to solve the design problem.

Fig. 7-10. Sketches show what a designer has in mind. (Ohio Art Co.)

These early sketches are often called rough sketches or roughs. This does not mean that they are crude and hard to understand. The name suggests only that the ideas are in their rough stage. They are not well thought out. The designer is looking for quantity (many ideas) rather than quality (the best idea).

Another reason for rough sketches is to solve specific design problems. Think back to the toy design problem. The designer must develop general sketches. The size and shape of the trucks must be developed, Fig. 7-11.

However, the designer explores specific design features too. For example, the wheel-axle assembly must be designed. Also needed is a method of attaching the wheels and axles to the truck body, Fig. 7-12.

Rough sketching techniques

Rough sketches try to show how the product will look. Therefore, designers use pictorial sketches. A pictorial is a drawing that shows an object as the eye sees it.

Fig. 7-11. Rough sketches show basic design ideas, not details.

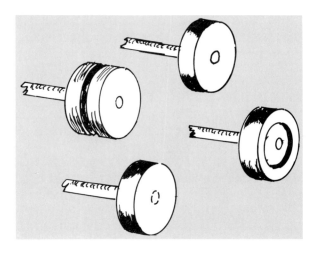

Fig. 7-12. Rough sketches can be used to develop specific design features.

One of two major types of pictorial sketches are used for rough sketches. These are:
1. Oblique sketches.
2. Isometric sketches.

Oblique Shapes

Oblique sketches, as shown in Fig. 7-13, show the front view directly. The designer

Fig. 7-13. Oblique sketches and drawings show the front view in its true size and shape.

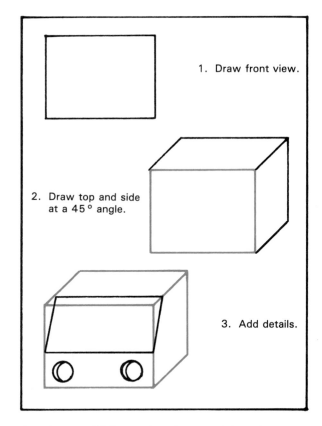

1. Draw front view.

2. Draw top and side at a 45° angle.

3. Add details.

Fig. 7-14. Oblique sketches and drawings are fairly easy to draw.

sketches this view first. Then the side and top view are developed using 45 degree lines. (The lines extend from the front view at an angle of 45 degrees.)

Oblique sketches are easy to develop. First, draw a box into which the object will fit, Fig. 7-14. Then add lines to show the outside shape. These may be angles or curves along the edges of the object. Finally, put in any interior details. These details show holes or other cavities in the object.

Oblique drawings have a major disadvantage. They tend to make the object look deeper than it is.

Isometric Shapes

Isometric drawings, as shown in Fig. 7-15, overcome this disadvantage. But they are harder to draw.

To make an isometric drawing, the designer first draws a vertical line, Fig. 7-16. This line is as long as the object is high. Then lines are drawn at 30 degrees away from the first line.

Fig. 7-15. Isometric sketches and drawings show objects more like the eye sees real things.

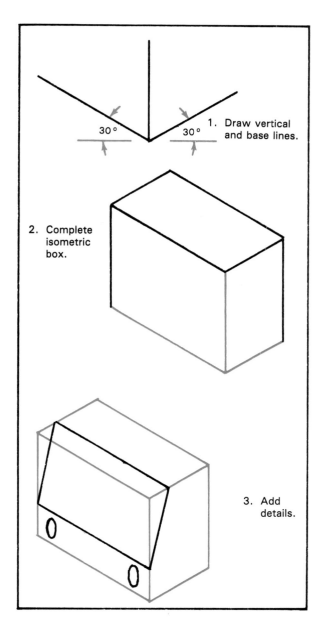

1. Draw vertical and base lines.

2. Complete isometric box.

3. Add details.

Fig. 7-16. Isometric sketches and drawings are drawn from one edge with the sides extending back at 30 degrees.

The left line is equal to the length of the object. The line to the right is the same as the width.

Next the designer draws the outline of a box using parallel lines. Then external details are added. Angles and curves are drawn to show the object's shape. Finally, internal holes and other details are added. Round holes appear as ellipses while square openings are trapezoids.

Refined sketches

Now the designer has produced a large number of rough sketches. These sketches probably contain many good ideas. But these ideas are not developed. The designer must select and improve the most promising ideas.

To do this, he or she draws refined sketches, Fig. 7-17. They clear up and extend the ideas suggested in rough sketches. A refined sketch may center on one rough sketch. However, refined sketches may combine ideas from several rough sketches.

The purpose of the refined sketch is to narrow down and improve the possible solutions shown on the rough sketches.

As with roughs, refined sketches are generally pictorial. They may use either oblique or isometric sketching techniques.

Selecting Promising Solutions

At several points in the design process someone must make decisions. Designs must be studied. Someone must give approval to continue developing the ideas.

Fig. 7-17. Refined sketches are used to develop ideas shown on rough sketches.

These decisions must be made carefully. A wrong choice can be costly. Also, no one wants to waste resources on products people neither need nor want.

Generally, several managers and product experts review the designs, Fig. 7-18. They study the product definition and the refined sketches.

Fig. 7-18. These designers are reviewing a possible toy design.   (Ohio Art Co.)

Their job is to screen the ideas. They will approve promising ideas for further development. Other ideas may be scrapped or saved for future development.

Often during the selection process, suggestions for design improvement are developed. These suggestions will be worked into the refined sketches. A common method for developing these suggestions is **brainstorming.**

Brainstorming is a group process used to solve a problem. It involves:
1. Gathering a group of people together.
2. Assigning them a problem to solve.
3. Encouraging them to think of as many solutions as possible.
4. Discouraging early criticism or evaluation of ideas.
5. Asking the group to finally expand, combine, and add to the ideas developed.

Results of a product design improvement brainstorning session will contain many ideas.

Each suggestion for improvement must be carefully reviewed. Some ideas are used to improve the design. These will be incorporated into the refined sketches. Other ideas will be rejected. They are considered inappropriate for some reason.

Refining Selected Ideas

The ideas that pass the initial management screen must be further developed. The ideas will be further refined and brought into reality. They will move off the designer's sketch paper and into solid form. This task often involves at least three major activities:
1. Developing **detailed sketches.**
2. Developing **models.**
3. Preparing **renderings.**

Detailed sketches

Look back at the rough sketch in Fig. 7-11. How big is the object? No one can tell. The design must be sized. People must know how big it will be.

To show size, detailed or dimensioned sketches are made, Fig. 7-19. These sketches

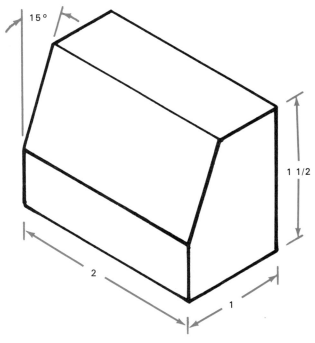

Fig. 7-19. Detail sketches show sizes for the product idea.

show three types of information:

1. Size information—length, width, diameter, etc.
2. Location information—distance from one or more points.
3. Geometric information—round, square, angled, etc.

The detailed sketches are necessary for the next design step. This step is modelmaking.

Modelmaking

So far all design work has been on paper. Two-dimensional sketches show how the object will look. But all objects are three dimensional. They have depth as well as height and width.

Sketches cannot fully show how an object will look. Neither will sketches tell us how well a design will work. Therefore, at some point, three-dimensional models must be made. These models may be either **mock-ups** or **prototypes.**

Mock-ups are appearance models, Fig. 7-20. They show how the object will look in real life. Most mock-ups are made from easily worked materials. Clay, wood, and cardboard are good mock-up materials.

Prototypes, as shown in Fig. 7-21, are working models. Their purpose is to test the operation of the object. They are usually made of the same materials that will be used for the actual product. Prototypes are often subjected to testing. This insures that the product will be safe and operate well.

The development of fast computers with vast memories has led to a new type of model. These, as shown in Fig. 7-22, are **computer**

Fig. 7-21. A prototype is a working model which may be used to test product performance. (Arvin Industries)

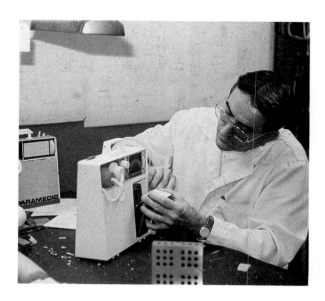

Fig. 7-20. Detailed sketches are used by modelmakers as they produce prototypes. (Ohio Art Co.)

Fig. 7-22. Products can be modeled on computers. (AMP, Inc.)

models. Computers can generate three-dimensional views on their color monitors. The engineers can then test and change its shape, if necessary. Also, complex computer programs can test the product without having an actual model.

The newest generation of passenger aircraft were designed and tested on the computer screen. They were "flown" on the computer before the first plane was built.

Renderings

Sometimes a fourth type of sketch is produced. This is a colored or shaded sketch called a rendering. This type tries to show final appearance of the object, Fig. 7-23. Renderings are often used to show products in which the appearance is of great importance. They are also widely used in package design and by architects.

Fig. 7-24. These children are helping a company test a new product.   (Ohio Art Co.)

Fig. 7-23. A rendering shows the designer's ideas for the appearance of the product.

Testing the Solution

The only good solution is one that meets the design problem. It must serve its purpose well.

The models that are produced must be tested. Physical items are subjected to actual conditions of use, Fig. 7-24. They are tested for many

characteristics. These include:
1. Function. Do they meet the stated need?
2. Performance. Do they work well?
3. Endurance. Will they last?
4. Operation. Can they be used easily?
5. Safety. Can they be used without harm to people?

The results of the testing activities may lead to design changes. Designs which perform poorly may be totally abandoned. Those designs which meet expectations move to the next design phase.

Selecting the Best Solution

All the design activities have led to a single event. A solution to the design problem must be selected. This solution must be the best overall. It may be a compromise between the "perfect" product and a product that people can afford to own. It must work well but be affordable. It must look good without being too expensive. It must be easy to build AND to repair. It must have value for the customer but make money for the manufacturer.

To arrive at this final solution, a group of managers and technical experts will review the designs, Fig. 7-25. They will study the models and test results.

They consider the proposals from several points of view. These include:

Fig. 7-25. These managers are participating in a product design review. (Ohio Art Co.)

1. Market. Will the product meet human needs and wants? Can it compete with other products on the market? Does the company know how to sell the product?
2. Production. Can the product be built economically? Does the company have the people and equipment needed to make the product? Does the product fit with other products being made by the company?
3. Finance. Can the company afford the initial investment required? Will the product make the company money?
4. Technical. Does the company have the engineering knowledge required to design and build the product?
5. Environment. Will the product harm the environment? Will it meet state and federal standards? Does it cause harmful pollution (air, water, sound)? Can the product be recycled or disposed of safely after use?
6. Social. Will the product fit within the society's social, cultural, and religious standards? Will the product damage the home or other societal institutions?

Successful proposals will be approved for engineering. The product idea will be prepared for production.

Interpreting the Solution

Approved designs must be changed into finished products. Parts must be made or pur-chased. These components must be assembled into products. All these activities require communications. People must know the exact size and shape of the parts. Location of parts within the product must be known. Also, the number and type of parts needed to build the product must be understood. In short, the design must be interpreted or communicated.

There are three common methods used for these tasks. These are:
1. Detail drawings.
2. Assembly drawings.
3. Bills of materials.

Detail drawings

Detail drawings provide the size and shape information for a single part. These drawings are used by manufacturing personnel to make the parts.

Typically detail drawings are orthographic projections. They show the part in one, two, or three views, Fig. 7-26. The drawings used are as follows:
1. One-view: for flat parts from a standard thickness material (sheet metal stampings, hardboard parts, etc.).
2. Two-view: for cylindrical parts (round shafts, discs, etc.).
3. Three-view: rectangular and irregular shaped parts.

Making orthographic drawings

An *orthographic* (multi-view) drawing is fairly complex to make. The steps usually followed are:
1. The surface with the most detail (holes, notches, etc.) is selected as the front view.
2. The front view is drawn in the lower left portion of the drawing.
3. Lines are drawn about 1 in. from and parallel to the top and right side of the front view.
4. A top view is drawn above the front view.
5. A right side view is drawn to the right of the front view.
6. Dimensions are added to show:
   a. The size of the part.
   b. Location of features (holes, notches, etc.).

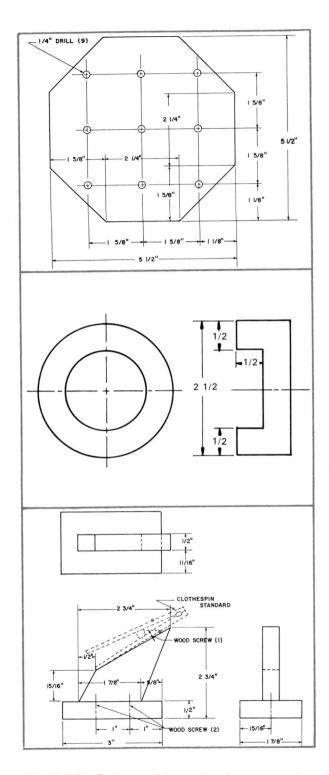

Fig. 7-26. Orthographic projections may be one-view, two-view, or three-view drawings.

Most products will require several detail drawings. One will be prepared for each different part.

Assembly drawings

Assembly drawings show workers how the parts fit together to make a product. They show the location of each part in an assembly or product. Sometimes products are shipped unassembled. The customer, then, uses an assembly drawing to put the product together.

Often, assembly drawings are exploded pictorials. They are isometric or oblique drawings in which the parts are pulled apart. Fig. 7-27 is a sample of an exploded assembly drawing.

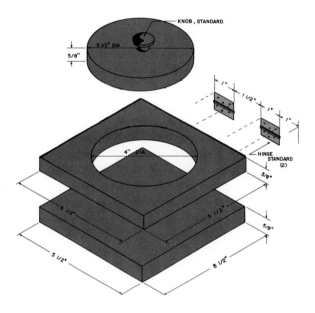

Fig. 7-27. An assembly drawing shows how parts of a product fit together.

Bill of materials

The final major design communication device is a bill of materials. This document lists all the parts that are needed to make one product, Fig. 7-28. It shows the part names and numbers, number of parts needed, part size, and the material to be used. Generally the parts are listed first. Purchased parts and hardware are listed last.

COMPLETE THE FOLLOWING BILL OF MATERIALS FOR THE SELECTED PRODUCT

PART NO.	NUMBER NEEDED	PART NAME	SIZE					MATERIALS
			T	X	W	X	L	
G101	1	BASE*	3/4		4 1/2		15 3/4	PINE OR REDWOOD
G102	1	SMALL END*	3/4		4 1/2		2	PINE OR REDWOOD
G103	1	LARGE END*	3/4		4 1/2		3	PINE OR REDWOOD
G104	2	ROD	1/4				18	MILD STEEL
G105	1	BALL	1				1 1/4	WELDING ROD
	4	SCREWS	NO. 8				1 1/4	REJECT BALL BEARING
	2	ROD ASSEMBLY PINS	NO. 16					BRADS

Fig. 7-28. A bill of materials lists all parts and hardware needed to make a product.

Presenting the Solution for Approval

All products must have management approval before they are built. Often a formal presentation of a new product is made. Designers and engineers prepare written reports. They outline the product and the anticipated costs. Also the benefits to the company and the expected market reaction is presented. Profit projections (estimates) are generally included in the report.

In most cases the written report is followed by an oral report. The designers and engineers meet with management personnel. They present the product for approval. They use charts, slides, transparencies, and other media to support their presentation.

Successful presentations lead to releasing the product for production.

SUMMING UP

The process for designing products has nine steps. Similar processes are used in designing construction projects.

Total technological systems also follow a design process. The system is described, components are designed or selected, the system design is reviewed, and the system is constructed.

Using the design process helps reduce the number of product and system failures. It also helps insure that the products and systems meets human needs without seriously affecting the environment.

KEY WORDS

These words are used in this chapter. Do you know their meaning?

Assembly drawing, Bill of materials, Brainstorming, Computer model, Design, Detail drawing, Detailed sketch, Feedback, Mock-up, Model, Prototype, Refined sketch, Rendering, Rough sketch.

TEST YOUR KNOWLEDGE
Chapter 7

Do not write in this text. Place answers to test questions on a separate sheet.
1. Design is (select best answer):
   a. Making a drawing of a product.
   b. The actual plans for a device that solves a problem.
   c. All of the activity that goes into making a new product.
2. List and describe the design steps.

3. A drawing which just gives the outline and shape of a product is called a _____ sketch.
4. A sketch which shows a product as the eye would see it is called a:
   a. Refined sketch.
   b. Pictorial sketch.
   c. Working drawing.
   d. Rendering.
5. Working models are called prototypes. True or False?
6. Most mock-ups are made from _____ materials.
7. A _____ _____ _____ is a list of all of the parts needed to make one product.
8. Interpreting the solution through drawings, bills of materials, and specification sheets. ducing a new product. True or False?

ACTIVITIES

1. Develop a product definition for a product that you need.
2. Find and photocopy five designs which meet your product definition.
3. Prepare a set of rough and refined sketches for your product definition.
4. Participate in a brainstorming group which seeks solutions to a school problem such as reducing the lunch room lines.

People responsible for design must hold many meetings. There they discuss the merits of different designs. From such meetings come many improvements on original design concepts. (Chicago Tribune)

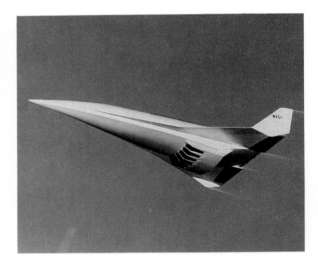

Sometimes a new idea for a product is drawn up by an artist so others can see how it might look. This is an artist's drawing of a new aerospace plane. (NASA)

APPLYING YOUR KNOWLEDGE

Everyday, new products and technological systems are designed. People are creating things to make our lives easier and better. They are completing a series of activities including:

1. Defining design problems.
2. Gathering information about the design problem.
3. Developing several solutions to the problems.
4. Selecting the best solutions.
5. Refining the selected design solutions.
6. Testing the solutions.
7. Selecting the best solution that passed the tests.
8. Interpreting the solution through drawings, bills of materials, and specification sheets.
9. Presenting the solutions for approval.

In this activity you will complete some of these steps. A design problem has been developed for your use. It is:

Design a greeting card for (holiday or event) which will communicate to (boys, girls, men, women) ages (specify).

You and your teacher will need to complete the problem statement by selecting an event and an audience. An example of a completed design problem follows:

Design a Thanksgiving Day greeting card which is appropriate for boys and girls 9 to 13 years old.

Equipment and Supplies

8 1/2" x 11" paper
Pencil
Ruler
Clip art (sheets of line art and pictures)
Drawing templates (circle, ellipse, etc.)
Colored pencils or markers
Lettering templates, typewriter, transfer type, etc.
Colored spirit duplicator masters (optional)
Spirit duplicator (optional)

Procedure

Your teacher will divide you into groups of two to four students as a greeting design team or ask you to design a card yourself. Each individual or group will complete the following design steps.

Defining the problem:

1. Obtain or develop a complete design assignment for the greeting card.

Gathering information:

2. Measure a group of cards that your teacher has for you. Determine common sizes for:
   A. Standard cards.
   B. Contemporary (studio) cards. See Fig. 7A for examples.

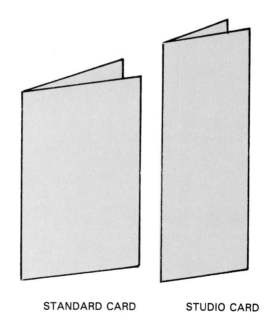

STANDARD CARD   STUDIO CARD

Fig. 7A. Types of cards.

3. Participate in a survey in the class to find out the type of cards your classmates like. This survey will ask which of the following card themes and styles each person likes:
   A. Card theme:
      1. Serious.
      2. Funny.
   B. Card size:
      1. Standard (about 4 1/4" x 5 1/2").
      2. Studio (3 1/2" x 8 1/2").
   C. Card message:
      1. Says something nice.
      2. Funny message directed at the person receiving the card.
      3. Funny message not directed at the person receiving the card.
4. For each survey question record:
   A. Number of people choosing each category. See Fig. 7B.
   B. Percentage of people choosing each category. See Fig. 7B.

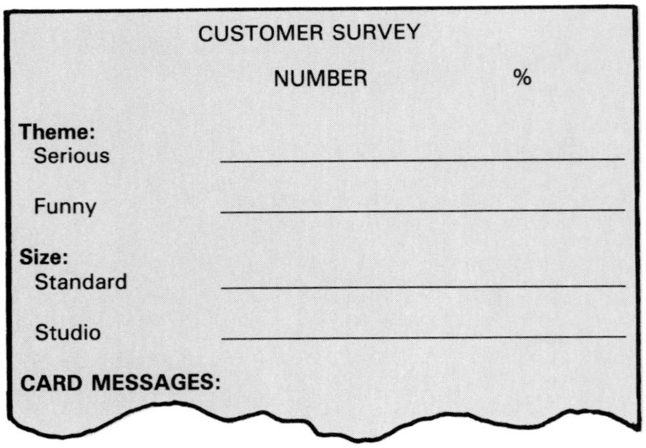

CUSTOMER SURVEY

NUMBER   %

Theme:
Serious _____
Funny _____
Size:
Standard _____
Studio _____
CARD MESSAGES:

Fig. 7B. Sample of a survey form.

Redefining the problem

5. Write a new program statement using:
   A. The original design problem.
   B. The information gathered in class.
   An example of an enlarged design statement:

   "Design a funny studio-sized Thanksgiving greeting card especially for the 9 to 13 year old boy or girl who receives the card."

Developing and selecting solutions — message:

6. Write drafts of three to five messages (copy) for a card which will meet your design problem, Fig. 7C.
7. Share your message with another design group or with a group of four or five individual designers. Encourage the group to suggest ways to improve the message.
8. Select your best two messages.

MESSAGE DRAFTS

THIS IS THE DAY
TO BE
THANKFUL
WE'RE NOT TURKEYS

Fig. 7C. Sample draft of a message.

Developing and selecting solutions — graphics (art):

9. Develop rough sketches of three to five illustrations to support each message. These may be freehand drawings or clip art, magazine pictures, etc. See Fig. 7D.
10. Present your sketches to the same group that reviewed your message designs. Have the group react to the way the messages and the drawings go together.
11. Select the best sketches for the card development.

Fig. 7D. Develop rough sketches of three to six card designs.

Refining the selected solutions: _____

12. Using the suggestions you received from your classmates:

    A. Refine your two best messages.

    B. Make refined sketch of the drawings selected for each message.

       You now have two messages with a different illustration for each.

13. Prepare two prototype cards for each refined message/illustration set, Fig. 7E. This will require you or your group to develop four prototype cards. Card "A" and card "B" will have the same message and illustration but will have different layouts (arrangement of the illustration and the message). Likewise, cards "C" and "D" will have different layouts for your second illustration and message set.

Testing and selecting the best solution: ___

14. Present your four cards to prospective customers (members of the class, a group of adults, etc.)

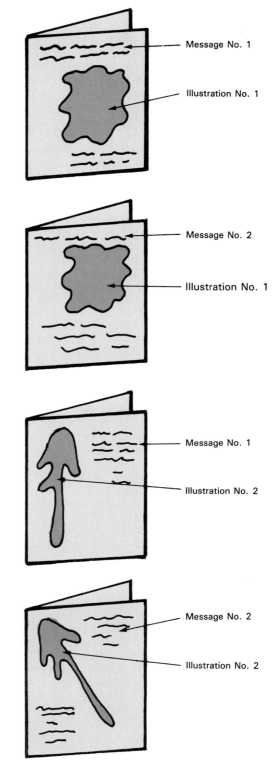

Fig. 7E. Prototype cards.

15. Have the group carefully study each card then rank the one they like best No. 1; their second choice No. 2, etc.
16. Using the customer ranking, select your best card design.

Interpreting the solution: _____
17. Make a final layout of the selected card, Fig. 7F. This layout should include:
    a. Color of paper to be used.
    b. Size and style of type to be used.

c. Color of ink to be used on the type and the illustrations.
d. Location of illustrations and type.
e. Method of folding the paper to make the card.

Producing the card (optional) _____
18. Using colored spirit duplicator masters, prepare a duplicator master for the card.
19. Produce a card for each member of the group.

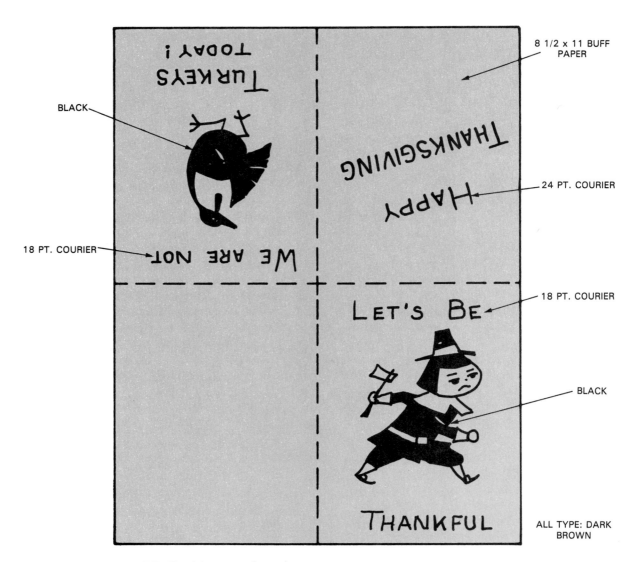

Fig. 7F. Final layout of card.

Chapter 8
Manufacturing Systems

Everyone uses manufactured products daily. We ride in manufactured vehicles. We buy records, tapes, and compact discs which were manufactured. We put on manufactured clothes. We enter buildings through manufactured doors. We read newspapers produced on manufactured printing presses. With manufactured pencils we write on manufactured paper. Manufactured games, sporting goods, and toys entertain us. See Fig. 8-1.

This world would be very different without manufactured products. Each of us needs the output of manufacturing systems.

MANUFACTURING

Manufacturing changes raw materials into useful products. It produces goods inside a fac-

tory. These goods are then shipped to stores. There, customers buy them to meet their needs and wants.

Manufacturing generally includes two major steps, Fig. 8-2. The first converts raw material into industrial materials. Trees are made into lumber, plywood, and paper. Ores are converted into sheets of metal. Natural gas is converted into plastics. Glass is made from silica sand. These processes were briefly described in Chapter 3. They are called primary processes.

Fig. 8-1. These are but a few of the toys manufactured by one company. (Ohio Art Co.)

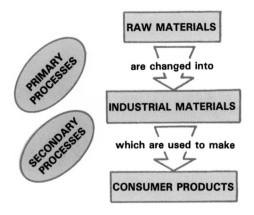

Fig. 8-2. Manufacturing includes primary and secondary processes.

Fig. 8-3 shows some typical **primary processing** activities.

The other type of manufacturing is called **secondary processing.** These processes change industrial materials into usable products. Through secondary processing, plywood becomes furniture. Sheets of metal become household appliances. Glass becomes bottles and jars. Plastics becomes dishes.

Primary Processes

Primary processes include three major groups. These were discussed in Chapter 3. The following briefly reviews what you learned there:
1. Mechanical processing: cutting, grinding, or crushing the material to produce a new form. These processes include cutting lumber and veneer, making cement, crushing rock into gravel, and grinding wheat to make flour.
2. Thermal processing: using heat to change the form or composition of materials. These processes include smelting metallic ores (making copper, steel, etc.) and fusing silica sand into glass.
3. Chemical (and electrochemical) processing: using chemical actions to change resources into new materials. These processes include refining aluminum from bauxite and producing most plastics from fossil fuels.

The output of primary processing is called standard stock. The materials are available in standard sizes. Most plywood is produced in 4 by 8 ft. sheets. Sheet metal is often sold in 24 x 96 in. sheets. Plastics are produced in standard pellets. Lumber is sold in a number of standard sizes. Many of us have heard people talk about 2 by 4s. This is a standard lumber size for the construction industry.

Standard stock must be further processed before it is useful. A sheet of plywood is of little value to a person. It becomes useful when it is made into a desk or dog house, Fig. 8-4. Likewise, few of us have use for plastic pellets. But, they can be converted into automobile trim, bowls, and fabric. Then the plastic takes on value to most people.

Secondary Processes

Secondary processing activities can be grouped under six headings: casting and molding, forming, separating, conditioning, assembling, and finishing. See Fig. 8-5.

Casting and molding

The first three secondary processing activities (casting and molding, forming, and separating) give materials specific size and shape. In one type of process, the material is first made a liquid. The liquid is poured or forced into a *mold.* (This is a cavity in the shape of the part or product.) There the material hardens. This process is called casting or molding, Fig. 8-6.

All casting and molding activities follow some common steps:
1. A mold is produced.
2. The material is made liquid.
3. The material is put into the mold.
4. The material hardens.
5. The finished item is removed from the mold.

If you have ever made a "home-made" popsicle, you have produced a casting. The first step was to get a mold. This mold had a cavity in it. The shape of the cavity gives the finished product its shape.

Some casting processes use a new mold for each product. These molds are usually made of sand or plaster. These processes are called one-shot or expendable mold casting.

Fig. 8-3. Converting trees into lumber involves a number of steps. A—Felling mature trees. B—Moving logs to the mill. C—Removing the bark from the logs. D—Cutting logs into slabs. E—Cutting slabs into boards of standard widths and lengths. (Weyerhaeuser Co.)

Fig. 8-4. Lumber and plywood take on added value when they are used to build homes and furniture. (Weyerhaeuser Co.)

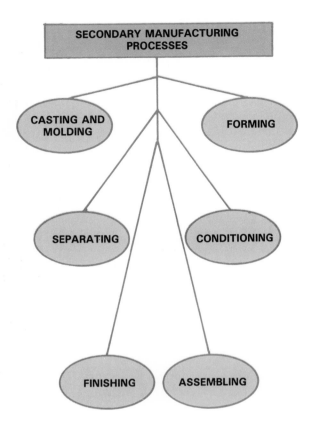

Fig. 8-5. Secondary manufacturing processes.

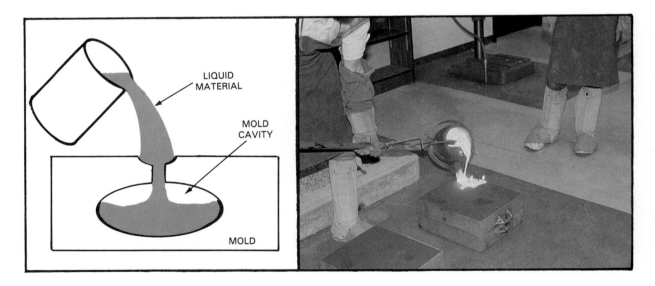

Fig. 8-6. Casting and molding processes. Liquid material is poured into a mold (cavity) where it hardens.

Other casting processes use metal or plaster of paris (a ceramic cement made from gypsum) molds. These molds can be used a number of times. This type of casting is called permanent mold casting.

Most materials that are cast are solids. They must be made liquid. Often they are melted. Other materials, such as clays, are suspended in (mixed with) water.

The liquid material is then put into the mold. In some cases, the material is poured. Gravity sometimes draws the material into the mold cavity. In other cases, machines force the material into the cavity.

Once in the cavity, the material must solidify. Sometimes it cools to become hard. With other materials, water is absorbed by the mold to create a solid part.

Finally the hardened material must be removed from the mold. One shot molds are broken away from the cast part. Permanent molds are opened. The finished casting is ejected from the mold.

Many metallic, plastic, and ceramic materials can be formed by casting processes. Typical examples are automobile engine blocks, parts for plastic models, ceramic bathtubs and lavatories, and plaster wall decorations.

Forming

Sometimes industrial materials are shaped using force. The materials are squeezed or stretched into the desired shape. These processes are called **forming**, Fig. 8-7.

All forming processes require two things. They must have:
1. A shaping device.
2. A force.

One forming process uses a shaping device called a **die**. This die is usually a set of metal blocks. Cavities are machined in them. Hot material is placed between the die halves. A hammer or press applies force. One die half moves toward the other half. As they close, the dies apply force on (squeeze) the material. This causes the material to flow into the die cavities. See Fig. 8-8.

This process is called **forging**. It is used to make hand tools, automotive parts, and other products requiring high strength.

Another forming process uses a single shaped die or mold. The material may be forced into the die cavity or it may be drawn around a mold. An example of this type of process is **thermoforming**. A plastic sheet is placed above a mold. The sheet is heated. It is then lowered onto the mold. A vacuum is pulled in the cavi-

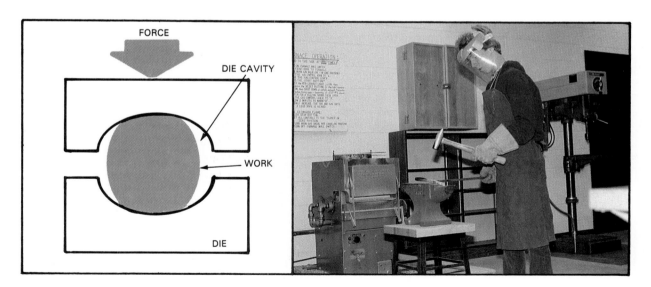

Fig. 8-7. Forming process. Left. Simplified drawing of a die. Right. Blacksmithing is a forming process.

Fig. 8-8. These workers are forming a metal part. Do you see the die halves in the center? (Fansteel, Inc.)

ty or around the mold. This causes the hot plastic to draw tightly to the sides of the mold. Plastic parts of all shapes are produced by thermoforming.

Rolls are also used to form materials. The material is fed between rotating rolls. This action stretches and squeezes the material into a new shape. This process, called roll forming, is used to make corrugated roofing and large tank parts.

These are but three forming processes. There are many more that use force and a shaping device to form materials.

Separating

Some manufacturing processes shape material by removing excess stock, Fig. 8-9. The extra material is cut, sheared, or burned away. These processes are called separation. They separate or remove the unwanted portion of the workpiece. This leaves a properly shaped part.

One type of separation is called **machining**, Fig. 8-10. These processes use a tool to cut chips of material from the workpiece. All machining requires motion between the tool and the workpiece.

The tool may spin to make the chip. Drilling and many sawing operations use a rotating (spinning) tool. In other cases the work is rotated against a solid tool. Lathes use this action. In still other cases the tool is drawn across the stationary (unmoving) work. The band saw and scroll (jig) saw move the tool across the work to make the cut.

A second separation action is called **shearing.** This process uses blades which move

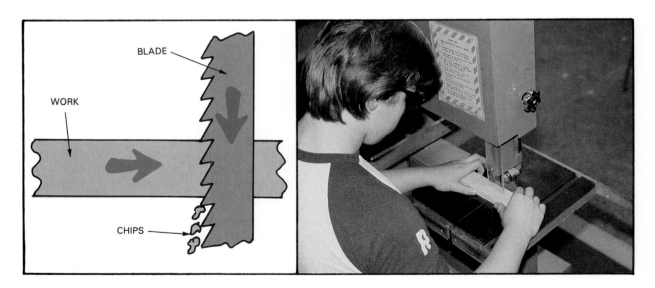

Fig. 8-9. Separating process. Saws, knives, or other tools cut away unwanted material.

Fig. 8-10. In this machining operation a jet engine part is being machined. (Fansteel, Inc.)

against each other. The material is placed between the blades. Then force is applied by the moving blades. This force fractures or breaks the material into two parts. Scissors, tin snips, and shears are shearing tools.

The third type of separation process is called **flame cutting.** Burning gases are used to melt away unwanted materials. Oxyacetylene cutting is an example of flame cutting.

Newer separating processes use beams of light, sound waves, electric sparks, and even jets of water to cut away unwanted materials. Laser machining, ultrasonic (high sound) machining, and electrical discharge machining are examples of this type of separating, Fig. 8-11.

Conditioning

A fourth type of secondary processing is **conditioning.** These processes alter the internal structure of the material. This action will change the properties of the material. Conditioning may be done with heat, mechanical forces, or chemical action, Fig. 8-12.

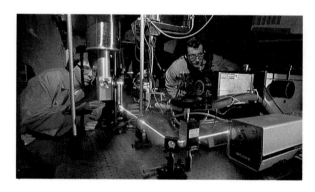

Fig. 8-11. These scientists are experimenting with a laser which could be used for cutting materials. (Amoco Corp.)

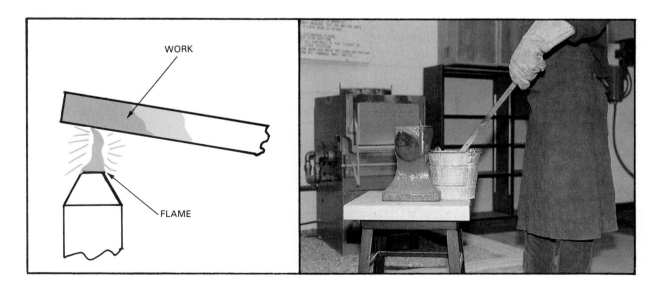

Fig. 8-12. Conditioning process. Metal is heated, then cooled, to make it harder.

The most common conditioning activity is thermal (heat) conditioning. The material being conditioned is heated to make it harder, softer, or easier to work.

Three major types of heat treatments are used on metals. These are:
1. Hardening.
2. Annealing (softening).
3. Tempering (removing internal stress).

Ceramic materials can also be conditioned by heat. This process is called **firing.** It involves slowly heating material to a very high temperature. The item is then allowed to cool slowly. During the process a glasslike ingredient in the ceramic material melts. This coats the clay particles in the material. As the material cools it become solid. The solid, glasslike materials bond the clay particles into a rigid structure. The result is a very hard, brittle product.

Assembling

Nails, screws, baseball bats and combs are all one-part products. But most products are made up of several parts. The act of putting the parts together is called **assembling.** Parts may be held together using mechanical fastening or bonding, Fig. 8-13.

Mechanical fasteners grip the parts and hold them in place. Typical mechanical fasteners are nails, screws, rivets, nuts and bolts, staples, and stitching (sewing).

Bonding permanently assembles parts together. This may be done by two major actions:
1. Fusion (welding).
2. Adhesive bonding.

Fusion uses cohesion. It uses the same forces that hold the molecules of the material together. The parts are melted at the joint (where they meet) area. This causes the materials to flow together. When the material cools a bond is formed. The two parts become as one. This assembling process is usually called welding, Fig. 8-14.

The second bonding technique uses an adhesive. (This is a sticky substance that holds the parts together.) An adhesive first must be able to attach itself to the parts. Then, an adhesive "bridge" is formed between the parts. They are glued together.

Different materials require different adhesives. A wide range of adhesives are available to adhere (stick) almost any two materials together. Plastic trim parts are adhered to automobile bodies. The skin of air-

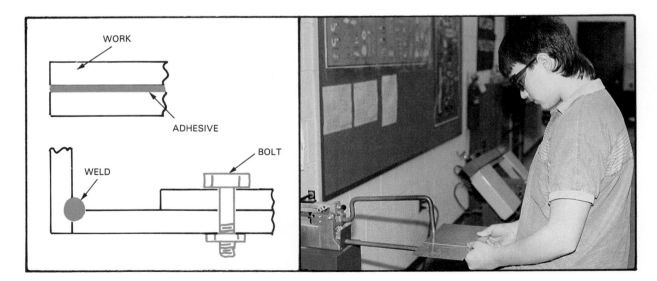

Fig. 8-13. Assembling process fastens parts together.

craft wings is adhered to the wing structure. Wallpaper is glued (adhered) to walls of homes and apartments.

Finishing

The final group of secondary manufacturing processes is finishing. This area includes all activities that protect and beautify the surface of a material, Fig. 8-15.

A finish is usually a surface coat applied to the material. This coating can be one of three basic types of materials:
1. Organic (plastic) material.
2. Metallic material.
3. Ceramic material.

Organic finishes are plastic materials suspended in a solvent. They are applied by brushing, spraying, rolling, or dip coating.

Fig. 8-14. The worker on the left is welding two parts together. The robot on the right is welding an air cooler together. (Coachman Industries and Arvin Industries)

Fig. 8-15. A finishing process. Paint can be sprayed onto a product. It protects the product and improves its appearance.

The finish dries when the solvent (thinner) evaporates. As this happens, the plastic material changes into more complex molecules. This new form produces a hard, uniform coat. The coat keeps water, oil, and other environmental elements from the base material. Look at Fig. 8-16.

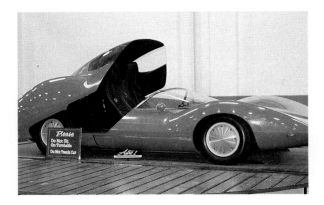

Fig. 8-16. The finish on this prototype automobile both protects and beautifies its surface.

Organic finishes are called paints, enamels, varnishes, and lacquers. They provide an attractive, protective coating for metals and woods.

Metals can be applied as a finish. In this process a base material is coated with the metal. This may be done several ways. The most common are:
1. Electroplating—using electricity to deposit the metal on the part.
2. Dipping—suspending the part in a vat of molten metal.

Chromium is a common metal coating material. The metal protects and provides a shiny surface. It is generally applied by electroplating the part.

Several other metals are applied by dip coating. Zinc coatings protect steel barn siding and roofing, garbage cans, and other steel products. The steel is dipped into molten zinc then allowed to cool. This process is called galvanizing.

A similar process applies tin to steel. The result is a tin-coated steel sheet. This material is widely used to make "tin cans."

Ceramic materials also make good coatings. Porcelain and glaze (glasslike material) are often used. The coating material is applied to the part cold. It is then heated to melt and fuse the finish to the product.

Many ceramic products are finished with glaze. The material provides a colorful and water-resistant coating for dishes, planters, and other products.

Porcelain is often used to coat metals and ceramic products. Some kitchen appliances are coated with porcelain enamel. Many bathroom fixtures also have a porcelain coating.

MANUFACTURING SYSTEMS

Secondary manufacturing processes are used to make products for everyday use. But they must be organized to be effective. They must be used in a manufacturing system.

There are four major types of manufacturing systems. These, as shown in Fig. 8-17, are:
1. Custom manufacturing.
2. Intermittent and batch manufacturing.
3. Continuous manufacturing.
4. Flexible manufacturing.

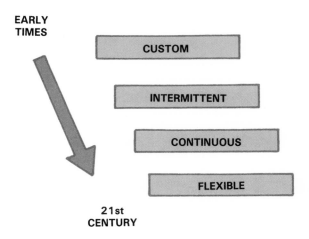

Fig. 8-17. The four major types of manufacturing.

Each of these systems is used today to make products. Each has advantages and disadvantages.

Custom Manufacturing

Custom manufacturing is the oldest system. In early history, one person made the entire product, Fig. 8-18. That person had all the skill needed to process the materials into products.

Fig. 8-18. This early machine allowed a worker to custom manufacture a gun stock.

Most products of colonial times were custom made. The silversmith made silver bowls and candle holders. The cobbler made shoes. The weaver made cloth. The tailor made clothing.

Later, custom manufacturing systems were used to make very special products. These products were designed for the customer. Only a few products were built to fill a need.

Today, spacecraft, ships, some cabinets and furniture, and clothing for special needs are custom manufactured. Sometimes, many people will work on the same product.

Intermittent Manufacturing

As the nation grew, custom manufacturing could not meet demand. There were a great many more people. They wanted more products. The skilled craftworker could not produce products fast enough.

Small factories were started. The products were made in small batches. Maybe a dozen or more candle sticks were made at a time. This system was called intermittent manufacture.

Intermittent manufacturing is widely used today, Fig. 8-19. In this system the parts for a product travel in a lot or batch. For example, suppose 100 bird houses are needed. One part is the front. First, workers select lumber to make the fronts. Next, they move the boards to a saw. Here 100 bird house fronts are cut to length. These parts are put in a tray. The tray moves to a drill press. Workers drill the entry hole for the bird in all 100 parts. The tray of parts travels to another drill press. Perch holes are drilled in each piece. The parts finally move to another saw. Here the roof peak is cut on all 100 parts. Do you see that the parts moved from operation to operation in a batch?

Fig. 8-19. The metal stampings in the boxes were produced in job lots. (Ohio Art Co.)

Continuous Manufacture

When many products are needed, continuous manufacture is generally used. The parts move down a manufacturing line, Fig. 8-20. At each station on the line a worker completes a specific operation. Workers at each station are trained to do the job quickly. The product takes shape as it moves along the line. Completed parts flow to an assembly line. Here the parts are put together to form the finished product.

Fig. 8-20. These workers are each completing one task in assembling a toy. (Ohio Art Co.)

Flexible Manufacturing

A new system of manufacturing is called flexible manufacturing, Fig. 8-21. It uses complex machines which are controlled by computers. Flexible manufacturing can produce small lots like intermittent manufacturing. But it uses continuous manufacturing actions. Thus, flexible manufacturing is seen as the way for the future. It produces low-cost products as they are needed.

DEVELOPING MANUFACTURING SYSTEMS

Manufacturing systems are developed for one purpose. They produce products to meet people's needs and wants. The development of

Fig. 8-21. This flexible manufacturing system produces parts for automobiles. (General Motors Corp.)

a manufacturing system involves several actions. These include:
1. Selecting operations needed to make the product.
2. Putting the operations in a logical order.
3. Selecting equipment to make the product.
4. Arranging the equipment for efficient use.
5. Designing special devices (tooling) to help build the product.
6. Developing ways to control product quality.
7. Testing the manufacturing system.

Each of these elements contributes to efficient production of products. They help us use technology wisely.

Selecting Operations

Most manufacturing systems design decisions are based on product drawings. These documents describe the product to be built.

One of the first system design steps is to decide which operations are needed. This may sound easy. A hole is needed. What could be simpler? But wait, should it be drilled, punched, cut with a laser, or produced by electrical discharge machining?

Each feature of the product is studied. Tasks to be performed are listed. Then methods for doing each task are selected.

The result of this activity can be a set of operation sheets. Each sheet lists all the operations needed to make a part.

Sequencing Operations

First the operations are selected. Then the planner puts them in the proper order. The product must be built efficiently. Also, moving the product from workstation to workstation must be considered. Inspections must be scheduled. Plans must be made for storing parts and product until needed or sold.

Remember our example of the bird house ends? The material for the end was first CUT to length. Then parts were MOVED to a drill press. A hole was DRILLED. Again the part MOVED to another drill press. There another hole was DRILLED. At this point, the part was INSPECTED. The quality and location of the holes need to be checked. Then the parts MOVED to another saw. There the gable (pointed) ends were CUT. The finished part was again INSPECTED. Finally, the parts MOVED to a STORAGE AREA. They wait for other parts so that assembly can start. This simple example includes:

1. Four operations (changing the shape or size of the material).
2. Four transportations (moving parts from station-to-station).
3. Two inspections (checks on the quality of the part).
4. One storage (placing the part in a safe place until needed).

All these words describe the order of operations, transportations, inspections, and storage acts. But sometimes words are hard to follow. Charts communicate better. The **flow process chart** shown in Fig. 8-22 contains the same information. These charts are often used to design new manufacturing activities. They are also used to study old procedures. The results can be a better way to make the part.

Selecting Equipment

Each operation requires equipment. Saws cut lumber. Drill presses drill holes.

For each manufacturing system, equipment must be provided. The Flow Process Chart helps engineers determine the equipment they need for the process.

PRODUCT NAME BIRD HOUSE – END	FLOW BEGINS STANDARD STOCK	FLOW ENDS FINISHED PART	DATE 10-17
PREPARED BY: R.T. WRIGHT SECTION: R & D		APPROVED BY: JCb	

PROCESS SYMBOLS AND NO. USED: OPERATIONS 4 · INSPECTIONS 2 · DELAYS 0 · TRANSPORTATIONS 5 · STORAGES 1

Task No.	Process Symbols	Description of Task	Machine Required	Tooling Required
		MOVE MATERIAL TO SAW	STOCK CART	
		CUT TO LENGTH	RADIAL SAW	STOP #301
		MOVE TO DRILL PRESS	CONVEYOR	
		DRILL LARGE HOLE	DRILL PRESS #1	DRILLING JIG 206
		MOVE TO DRILL PRESS	CONVEYOR	
		DRILL SMALL HOLE	DRILL PRESS #2	DRILLING JIG 306
		INSPECT		
		MOVE TO CIR. SAW	CONVEYOR	
		CUT GABLE	CIRCULAR SAW	SAWING FIXTURE
		INSPECT		
		MOVE TO STORAGE	STOCK CART	STORAGE TRAY
		STORE		

Fig. 8-22. Operation or flow process chart has been filled in for the sample birdhouse end.

Some equipment may already be owned. Other items will be purchased.

Arranging Equipment

The equipment must be arranged for production. Sometimes the machines will be used for a number of different products. Intermittent manufacturing activities will be used. Then, like equipment is grouped together. A roughing department may be formed to contain all saws, jointers, and surfacers. A finishing department may be equipped to coat several different products.

In other cases the equipment is set up to make only one product. A continuous manufacturing system is designed. Then the equipment is arranged to make the selected product. In our example, the line to make our bird house ends would contain the following equiment: a saw, then another saw, a drill press, then another drill press. One arrangement for this line is shown in Fig. 8-23.

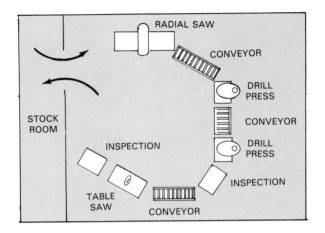

Fig. 8-23. A plant layout drawing for the bird-house end production line. See how handy both the beginning and the end of the line are to the stock room.

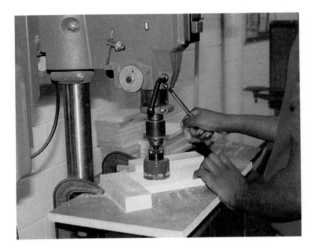

Fig. 8-24. This simple, student-built tooling insures that the hole is drilled in the same spot on all parts.

Designing Tooling

Many times, special devices make manufacture more efficient. These devices may hold a part so it can be machined, Fig. 8-24. Some may hold several parts for welding. Or they may be special dies for forming the material. All of these items are called **tooling.**

They are designed to make operation more efficient. They should make the operation:
1. Faster.
2. Easier to complete.
3. Safer.

For our bird house end, several pieces of tooling could be used. These might include devices to:
1. Hold the part so the entry hole is always drilled in the same place.
2. Hold the part so the perch hole is correctly located.
3. Hold the part so the gable ends are accurately cut.

Controlling Quality

Product quality is a major concern in manufacture. Everyone wants products that work well and look good. Therefore, parts and products must be checked for quality. This ac-

tion is called **inspection,** Fig. 8-25. The parts are compared with the drawings. Each part and product must meet the standards set by designers and engineers.

Parts failing the inspection may be scrapped. Other parts might be reworked. The defect may be removed from the part. Scrapping and reworking parts add cost to the product. Therefore, every effort is made to make the product right the first time.

Fig. 8-25. These technicians are testing electrical wiring components.   (AMP, Inc.)

Testing the System

The last step in manufacturing system design is testing the system. The parts of the system must be put in place. The machines must be positioned. Conveyors and other material handling devices must be installed. Tooling must be attached to the machines. Then the system is run.

People produce test products using the manufacturing system. This is called a **pilot run.** Engineers check to see that the operations are working correctly. They observe the flow of material. Also, product quality is carefully checked.

Pilot runs are important. They show where changes are needed. Results of a test run may show that:
1. Tooling needs to be improved.
2. Equipment should be shifted.
3. Workstations need to be relocated.

After the changes are made, engineers make another pilot run.

Changes continue until the system is operating properly. Only then will full-scale production start.

PRODUCING PRODUCTS

Manufacturing of products requires all the major types of resources. People use information and machines (tools), which are powered by energy, to change materials into products, Fig. 8-26.

Most manufacturing is managed. A group of people see that the system runs properly. These people are called managers. They make up management.

Managers do not make the products. They organize the systems so that the products are made efficiently. In doing this, they complete the four basic functions. They:
1. Plan—set goals.
2. Organize—divide tasks into jobs.
3. Direct—assign jobs and supervise workers.
4. Control—compare the results with the plan.

A good manager gets work done through other people. He or she provides direction and

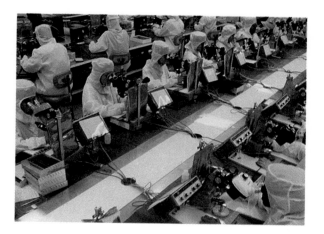

Fig. 8-26. Can you identify the resources being used in this semiconductor manufacturing line? (Harris Corp.)

support. Then workers can more easily make products.

MARKETING PRODUCTS

Today there are thousands of products available to each of us. These products must be marketed. Potential customers must be told about them. Generally, **advertising,** shown in Fig. 8-27, does this task. It tells about the product and its benefits.

Fig. 8-27. This two-page advertisement appeared in the three magazines shown. (Ohio Art Co.)

Ads are delivered in several ways. Some are placed in magazines and newspapers. Others are aired on radio and television. Still others are delivered through the mail or displayed on billboards and signs.

Advertisements make people act. They bring the customer to a store. There the sales effort takes place. Salespersons encourage the customer to buy the product, Fig. 8-28.

Expensive products, like automobiles, are sold by a salesperson. He or she presents the value of the product to the customer. Less expensive products (toothpaste, colas, etc.) are simply displayed. The customer actions are often based on the advertising effort, appearance of the package, or from previous experience.

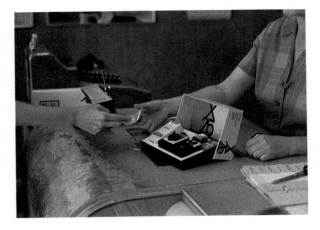

Fig. 8-28. The marketing effort persuades people to buy the product.

SUMMING UP

Manufacturing provides all the products we use. The food we eat, the clothes we wear, the vehicles we travel in were all manufactured.

To meet the large demand for products, complex manufacturing systems have been developed. They let people efficiently make things we need and want. Try to think of a world without manufactured goods. Not a happy thought, is it?

KEY WORDS

These words were used in this chapter. Do you know their meaning?

Advertising, Assembling, Casting and molding, Conditioning, Continuous manufacturing, Custom manufacturing, Die, Finishing, Firing, Flame cutting, Forging, Forming, Inspection, Intermittent manufacturing, Machining, Manufacturing, Pilot run, Primary processing, Quality, Secondary processing, Separating, Shearing, Thermoforming, Tooling, Welding.

ACTIVITIES

1. Select a simple product, then list the types of processes used to make it (casting and molding, forming, separating, conditioning, finishing, assembling).
2. Select a simple product, such as a book end, then list:
   a. The operations you would use to make it.
   b. The equipment or tools you think you would need.
   c. The points you would check for quality.
   d. Any special tooling you think you would need.
3. Suppose you were to drill a hole in the center of a 4 by 4 in. piece of wood. Sketch the piece of tooling you would use to make that hole accurately in 100 parts.

TEST YOUR KNOWLEDGE
Chapter 8

Do not write in this text. Place answers to test questions on a separate sheet.
1. Manufacturing changes _____ _____ into useful _____.
2. There are two types of manufacturing:
   a. Changing raw materials into industrial materials is called _____ _____.
   b. When industrial materials are changed into usable products it is known as _____ _____.

MATCHING TEST: Match the terms with the descriptions:

3. ___Cuts, grinds, or crushes material.
   A. Chemical Processing.

4. ___Heat materials to change them.
   B. Mechanical Processing.

5. ___Uses chemicals to make new materials.
   C. Thermal Processing.

6. Name the six secondary manufacturing processes.

7. Name the processes described:
   a. Uses a hollow form to shape a liquid material: _____
   b. Materials are forced or squeezed into a new shape: _____
   c. Excess stock is cut away by burning, shearing, or cutting: _____
   d. Alters the interior structure of the material: _____
   e. Fastening parts together by any means: _____
   f. Protecting or beautifying the outside of the product: _____

8. Read the description and name each type of manufacturing system:
   a. Products are made to order, one at a time: _____
   b. A bunch of a product is made at one time: _____
   c. Products made as they move down an assembly line: _____
   d. Made in small lots but uses assembly line and computer: _____

9. Deciding how a product will move through production is called _____ operations.

10. Flow process charts are sometimes used to show the order of operations. True or False?

11. Tooling is (select all correct answers):
    a. A device that holds a part while it is being manufactured.
    b. Designed to make manufacturing more efficient.
    c. Sometimes a special shape for forming the material.

12. Finding customers and telling them about the new product is known as _____.

Manufacturing required many persons qualified to operate machine tools. This scene is from a national skills competition sponsored annually for VICA. The competition promotes skill development among American youth.

Manufacturers have become very concerned that their products perform well. Workers are encouraged to produce quality work.
(Ohio Art Co.)

APPLYING YOUR KNOWLEDGE

Manufacturing is one of the four technological systems. The others are communication, construction, and transportation.

Manufacturing changes the form of materials to make them worth more. This may be done using custom techniques. People can make a product all by themselves. They can make it for their own use. More often a manufacturing system called continuous manufacture is used. The product is made on a manufacturing line.

In this activity you will use such a line. Working on the line, you and your classmates will change the form of materials. The original form will be strips of wood. The new, more valuable form will be a cassette holder.

During the activity, notice the use of the resources discussed in Chapters 2 through 6. See if you can see tools, materials, energy, information, and people being used. Also, note that most people will think the cassette holder is worth more than the strips of wood and nails used to make it.

Equipment and Supplies

Wood strips
   1/4" x 3/4" x random length (side strip stock)
   1/4" x 1 3/8" x random length (bottom strip stock)
   3/4" x 2 1/2" x random length (end stock)
5/8" x 18 wire brads
Back saws and/or band saw
Scroll saw
Hammer
Brace or drill press
3/4 auger or speed bit
Disc or oscillating sander
Abrasive paper
Sanding block
Special tooling:
   End cutoff jig

Side strip cutoff jig
Bottom strip cutoff jig
End hole drilling jig

Product Drawing and Bill of Material

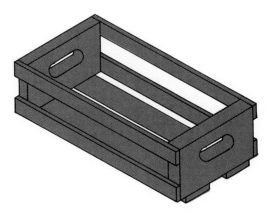

Fig. 8A. Isometric drawing of the cassette holder.

QTY	DESCRIPTION	SIZE	MATERIAL
2	Ends	3/4" x 2 5/8" x 4 3/8"	Pine or spruce
4	Side strips	1/4" x 3/4" x 9 3/4"	Pine or spruce
2	Bottom strips	1/4" x 1 3/8" x 9 3/4"	Pine or spruce
24	Brads	5/8 x 18	

Fig. 8B. Bill of materials for the cassette holder.

Procedure

1. Study the operation process chart, Fig. 8C, to determine the steps needed to make the cassette holder.
2. Read the explanation of the major steps to determine how to complete each step.

Operation Directions:

Oper. No.	Description
E-1	Set up a band saw or a miter saw to cut a 4 3/8 in. piece from the end

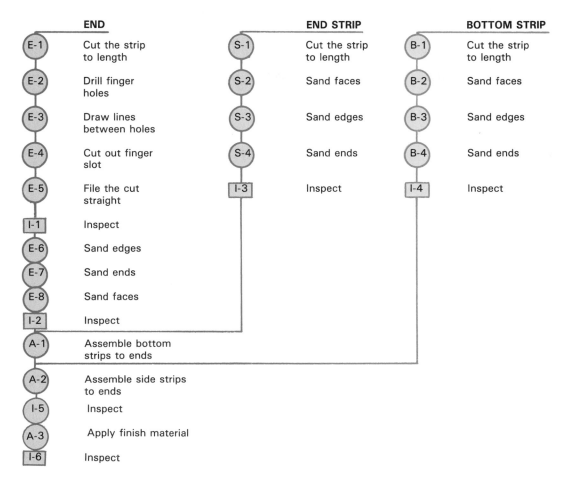

	END			**END STRIP**			**BOTTOM STRIP**
E-1	Cut the strip to length		S-1	Cut the strip to length		B-1	Cut the strip to length
E-2	Drill finger holes		S-2	Sand faces		B-2	Sand faces
E-3	Draw lines between holes		S-3	Sand edges		B-3	Sand edges
E-4	Cut out finger slot		S-4	Sand ends		B-4	Sand ends
E-5	File the cut straight		I-3	Inspect		I-4	Inspect
I-1	Inspect						
E-6	Sand edges						
E-7	Sand ends						
E-8	Sand faces						
I-2	Inspect						
A-1	Assemble bottom strips to ends						
A-2	Assemble side strips to ends						
I-5	Inspect						
A-3	Apply finish material						
I-6	Inspect						

Fig. 8C. Operation process chart. How many steps are there to produce the cassette holder?

stock. A special cutoff jig can be built to guide a backsaw during the cutoff operation.

E-2 Drill two 3/4 in. holes in the end according to the dimensions on the drawing in Fig. 8D.

E-3 Draw a line that connects the top of the two holes and another that connects the bottom of the holes.

E-4 Insert a scroll saw blade in one of the 3/4 in. holes. Tighten the blade in the saw. Cut out the marked section between the two holes.

E-5 Smooth and straighten the saw cuts with a file. CAUTION: Do not damage the curved portion at the ends of the slots.

I-1 Inspect the part for size and quality of the finger slot.

E-6 Sand the edges of the end parts on a disc or oscillating sander or with abrasive paper and a sanding block. CAUTION: Do not round the edges because the side strips must mount flat on these surfaces.

E-7 Sand the ends of the end parts on a disc or oscillating sander or with abrasive paper and a sanding block. CAUTION: Do not round the ends because the bottom strips must mount flat on these surfaces.

E-8 Sand the faces with an oscillating

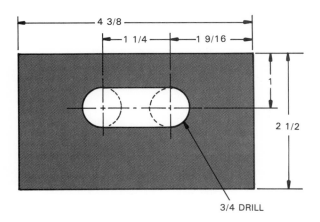

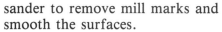

3/4 DRILL

Fig. 8D. Detail of finger holes needed for each end piece.

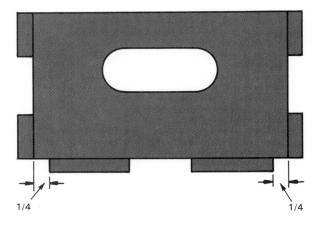

1/4   1/4

Fig. 8E. Locate botton strips and side strips as shown.

sander to remove mill marks and smooth the surfaces.

I-2 Inspect the results of the sanding steps. Look for square, flat edges and ends.

S-1; B-1 Set up a band saw or a miter saw to cut a 9 3/4 in. piece from the stock for the end strip (S-1) or the bottom strip (B-1). A special cutoff jig can be built to guide a backsaw during the cutoff operation.

S-2; B-2 Sand the faces with an oscillating sander to remove mill marks and smooth the surfaces. CAUTION: Be sure not to round these surfaces because one face must fit flat against the ends of the holder.

S-3; B-3 Sand the edges of the end parts on a disc or oscillating sander or with abrasive paper and a sanding block.

S-4; B-4 Sand the ends of the end parts on a disc or oscillating sander or with abrasive paper and a sanding block. CAUTION: Do not round the ends because they must be flush with the ends after assembly.

I-3; I-4 Inspect the results of the sanding steps. Look for square, flat edges and sides.

A-1 Locate the bottom strips as shown in Fig. 8E. Nail each end of each strip with two 5/8 in. by 18 wire brads. NOTE: A locating fixture may be used to speed this operation.

A-2 Locate the side strips as shown in Fig. 8E. Nail each end of each strip with two 5/8 in. by 18 wire brads. NOTE: A locating fixture may be used to speed this operation.

I-5 Inspect assembled product. Route any defective products to a rework station or to scrap.

A-3 Apply appropriate finishing material by brushing, wiping, or dipping.

I-6 Inspect final product.

Chapter 9
Construction Systems

The information given in this chapter will help you to:
- Define the term, construction.
- Describe how primitive humans sheltered themselves.
- List types of structures that are built on a site.
- Describe types of structures and discuss their uses.
- List and describe some construction materials.
- Define a construction system.
- Name a construction system's parts.
- Discuss how construction sites are selected and prepared for building.
- Discuss the tools and equipment used in construction.
- Describe or demonstrate how one type of structure is built.

Our neighborhoods are made up of many things that are constructed. We call these things structures. These are things made or put together where they are to be used, Fig. 9-1. Building these structures is called **construction.**

Construction is a series of carefully planned events. Construction technology uses materials, work, processes, and equipment to build a structure on a site. Management organizes these resources and uses them efficiently. (Efficiently means it is done without waste of time or resources.)

There was a time when people did no constructing. They made do with what nature offered. They had no knowledge of how to build. They also had no time! Tribes had to keep moving in search of food. Permanent shelter was of little use to them. The structures built reflected their life-style, Fig. 9-2.

PERMANENT STRUCTURES

In time, people began to settle down. They stayed in one place and grew food. Then there was a need for many kinds of structures. Look at Fig. 9-3.

They formed villages and built houses. They also built roads. This made travel from one place to another easier. Bridges were built over rivers. Other structures were built to store food.

The Egyptians, Greeks, and the Romans were the master builders among ancient people, Fig. 9-4. The Egyptians built the great pyraminds. The biggest one covered 13 acres. It was 480 ft. high. The pyramids were the tombs of kings.

The Greeks were famous for their beautiful temples. The Romans built many cities, roads, and bridges. Around 300 B.C. they built great stone waterways to pipe drinking water into

Fig. 9-1. Construction is an important technological activity. We construct tunnels, monuments, dams, bridges, roads and overpasses, water towers, factories, and houses.

B

A

Fig. 9-2. Hundreds of years ago humans built simple structures because they moved often. A—A Sioux dwelling of the western plains. B— A modern day copy. (Smithsonian Institution)

Fig. 9-3. Ancient Pueblo Indians constructed cliff dwellings in Arizona, New Mexico, Utah and southern Colorado. Built of limestone, they are grouped like apartment buildings. They were meant to protect the Indians from enemies.

A

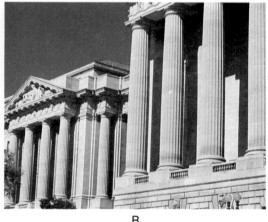

B

Fig. 9-4. The Egyptians, Greeks, and Romans were famous as ancient builders. Ruins of many of their structures can still be seen. A—The Egyptians built huge pyramids as tombs for their pharaohs (kings). This one covers 13 acres. (Walter Martz) B—Modern buildings often copy the architecture of Greek and Romans buildings.

their cities. These structures were called aqueducts. (Aqua means "water," and duco, "to carry.") See Fig. 9-5.

Fig. 9-5. The Appian Aqueduct was built to pipe water into Rome. The water channel is shown at the top of the pillars.

MODERN CONSTRUCTION

The structures of today meet many different needs. They can be grouped by their purpose. Included are:
1. Shelters—structures meant to provide protection for human activity.
   a. Residential buildings. These are places where people live.
   b. Commercial, public, or institutional buildings. These are structures for shopping, business, and schools.
   c. Civil structures. Structures that help move materials and people or give some other benefit to the public. They support many activities needed for living. Sometimes their only purpose is to provide beauty.

Residential Buildings

Shelter is one of our basic needs. It protects us from the weather. Homes make us more comfortable. They also give us conveniences. Can you think of things you would not be able to do if you did not live in a house or apartment?

Several different types of homes are built, Fig. 9-6. Single family dwellings stand alone

A

B

C

Fig. 9-6. People live in different kinds of structures. A—A single family dwelling. B—An apartment building houses several families. C—A residence is designed to shelter a family's activities.

on a plot of land. No other families live in the building. Condominiums and apartment buildings have several living units joined together. Sometimes the units are side by side. In this arrangement they share common walls. Apartment buildings are often multistoried (have more than one floor or level). A single, large building may have many separate dwelling units.

Commercial, Public and Industrial Buildings

Any structure that gives people protection, comfort, and convenience is a shelter, Fig. 9-7. Shopping centers, stores, schools, hospitals, office buildings, airport terminals, hotels, and industrial plants shelter certain activities. They are structures designed for people who shop, study, work, travel, or are ill. Can you think of any other shelters?

Civil Structures

Civil structures provide convenience or add beauty to the place where they are erected. "Civil" means that they are built for the community. Therefore, they are for the convenience of all the people who live in the community.

Civil structures include bridges, overpasses, roads, sidewalks, sewers, pipelines, utilities, ballparks, monuments, dams, and water towers. Civil structures are shown in Fig. 9-8.

Bridges and overpasses take people and vehicles over rivers, canals, and other obstacles. Overpasses cross obstacles such as other roadways or railroad tracks.

Sewers carry wastes away from homes and other buildings. Water towers store water for drinking, firefighting, or industrial and commercial activities. Pipes and cables which supply water, fuel, and electricity are called utilities. Perhaps you can think of other structures built for the convenience of people in your community.

Monuments are structures built to pay honor to people, ideas, and events. One of our most important monuments is the Statue of Liberty in New York. Another is the Lincoln memorial in Washington, D.C. Do you have monuments

A

B

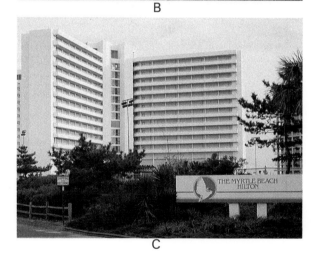

C

Fig. 9-7. A—The building shown shelters stores and restaurants. B—This structure houses offices for a company. (AMP Inc.) C—Hotels shelter travelers and vacationers.

Fig. 9-8. Airport terminals, sports arenas, and community water towers are civil structures. Do you know why?

in your community? Do you know what persons or ideas they honor?

CONSTRUCTION INPUTS

Construction technology needs many different inputs. It uses people, machines, and knowledge to create structures. The structures are created from many different materials and manufactured goods.

All the resources must be brought together in the right way at the exact time. If any resource is not there when it is needed the project could fail.

People

People are necessary to a building project. They perform many different tasks.

An architect will design the structure. An **engineer** will determine how much material is needed. An owner will provide the land and pay for the structure. A contractor hires and directs skilled workers, Fig. 9-9.

Estimators add up the costs of time, labor, and materials. They then make an accurate estimate of the building cost.

Project managers and supervisors help the contractor. They see that the work follows the specifications and codes (rules).

Carpenters, electricians, plumbers, and cement finishers shape the structure. Other skilled workers operate heavy construction equipment like cranes, bulldozers, and piledrivers, as shown in Fig. 9-10.

Information

Many kinds of information are needed. Plans are drawings of how a structure should be built. One kind of drawing is called a floor plan, Fig. 9-11. Drafters make plans for every part of the structure.

Materials and Manufactured Products

There are many different building materials and products, Fig. 9-12. Included are concrete, metal, lumber (wood), glass, and masonry units

Fig. 9-9. Contractors hire and direct work of construction workers. They also work with plans and with costs of construction. (Atlantic Richfield Co.)

Fig. 9-10. Many skilled people are needed in construction. A and B—Equipment operators fill a trench for telephone cable installation and dump fill around a foundation. C—Carpenter constructs structures of wood. (Disston Corp.) D—Form builders assemble a form that will be used to pour a pillar.

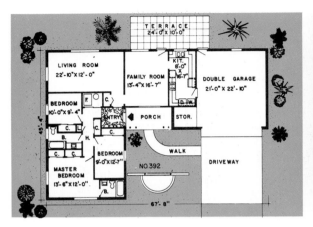

Fig. 9-11. Many drawings are needed to describe a structure. Left. A floor plan is used to show sizes and location of rooms in a house. Right. A rendering is a drawing that looks like a photograph.

Fig. 9-12. A few of the materials of construction. Left. Lumber and brick at a building site. Right. Doors at a lumber yard.

(masonry units are bricks and concrete block). Other materials used include stone, plastic laminates, gypsum board, plaster, fiberglass, and asphalt.

Other products are assembled before arriving at the construction site. Such products include doors, windows, and cabinets.

CONSTRUCTION PROCESSES

Building anything requires completing a series of steps. These have to be done in the right order. The steps are part of one technical process. It is called the construction process.

Construction projects almost always follow the same steps. The major steps are:
1. Planning, Fig. 9-13.
2. Constructing.
3. Servicing.

Planning

Planning begins with an idea. Someone decides to build. Every structure must fill a need or wish. A new highway may be needed.

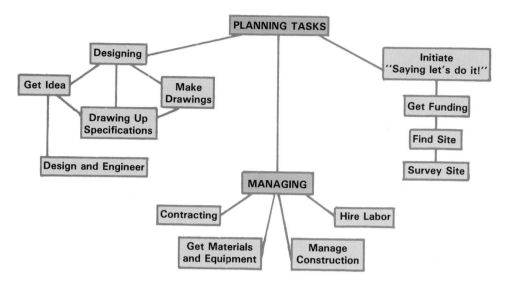

Fig. 9-13. There are many steps in planning construction projects.

It will provide safe, fast travel. A family wants a larger home. It answers their wish for more space. Some projects are built so the owners can sell them for a profit.

Getting funding

Building costs money. The next step in planning is to get funding. Money may have to be borrowed. Individuals pay for private projects. Public projects get funds from taxes.

Getting a site

Every structure requires land. Someone must locate and obtain a site (place to put the structure).

A dam or a bridge will be built on a river. There will be few choices of location.

A house or a factory is different. There are many places where they can be built. It is important to choose the right place. See Fig. 9-14. A house, for example, should be in a residential area. It should also be close to things that the owner needs. Being near schools, shopping, and jobs may be important. Also, the land must be for sale.

Buying the site

Suppose your friend has a bicycle you want to buy. First, you agree on a price. Then you give money for the bicycle. The whole exchange may take only a few minutes.

Buying land takes more time. The process also involves more steps. One way to get land is by negotiation.

Fig. 9-14. A building site. A sign shows that the owner is willing to sell land.

Negotiation

When land is bought for a private structure, the buyer and seller **negotiate.** They bargain until a selling price is agreed upon. Then they agree on what the seller can remove from the property. Buyer and seller also work out when the property will change hands. Then they sign a contract. This is a carefully worded document (official paper). It says the buyer will buy the property for an agreed price on a certain date. Fig. 9-15 shows a contract for the sale of real estate.

Condemnation

Condemnation is a procedure that governments may use to buy land for public use (roadway, school, courthouse, etc.). It is used when an owner does not want to sell. The government has a right to buy the land at a fair price. The owner is forced to sell for the public good.

Surveying the Site

When land is bought, the new owner must know where the property begins and ends. These are known as **boundaries.** They are

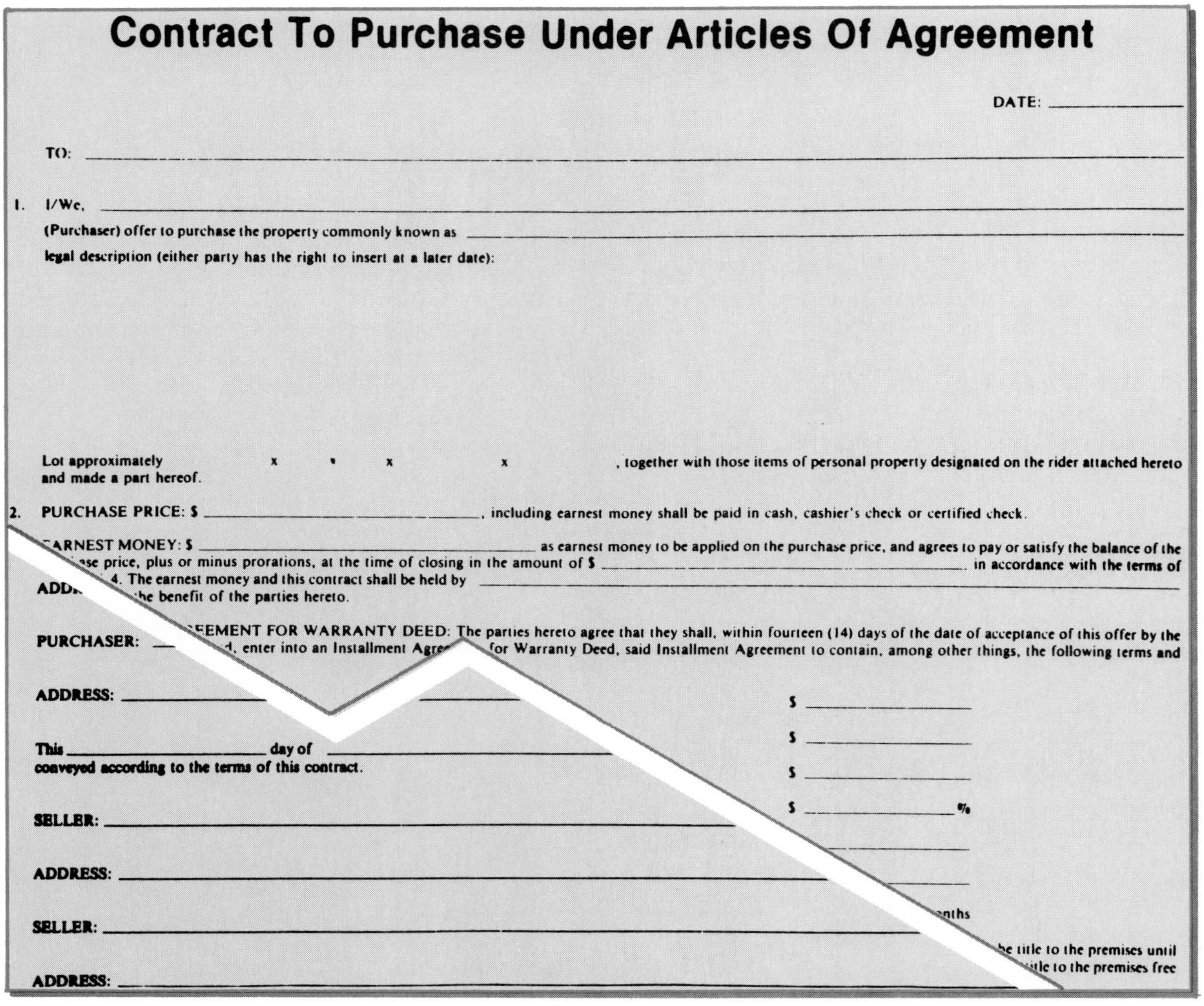

Contract To Purchase Under Articles Of Agreement

DATE: _____

TO: _____

1. I/We, _____

(Purchaser) offer to purchase the property commonly known as _____

legal description (either party has the right to insert at a later date):

Lot approximately   x   •   x   x   , together with those items of personal property designated on the rider attached hereto and made a part hereof.

2. PURCHASE PRICE: $ _____ , including earnest money shall be paid in cash, cashier's check or certified check.

EARNEST MONEY: $ _____ as earnest money to be applied on the purchase price, and agrees to pay or satisfy the balance of the price, plus or minus prorations, at the time of closing in the amount of $ _____ . in accordance with the terms of 4. The earnest money and this contract shall be held by _____ he benefit of the parties hereto.

PURCHASER: _____ EEMENT FOR WARRANTY DEED: The parties hereto agree that they shall, within fourteen (14) days of the date of acceptance of this offer by the ____ , enter into an Installment Agre____ for Warranty Deed, said Installment Agreement to contain, among other things, the following terms and

ADDRESS: _____

This _____ day of _____
conveyed according to the terms of this contract.

SELLER: _____

ADDRESS: _____

SELLER: _____

ADDRESS: _____

$ _____
$ _____
$ _____
$ _____ %

nths

he title to the premises until
itle to the premises free

Fig. 9-15. A contract for sale of a lot. A description of the property and its location is added to a standard form.

established by doing a survey, Fig. 9-16. Two common instruments are used for this purpose. They are the **transit** and a steel tape called a **chain.** The transit measures the angle of the boundary lines. The chain measures the distance along the boundary lines.

Soil testing

Another type of survey is the soil test. Builders need to know how well soil will sup-

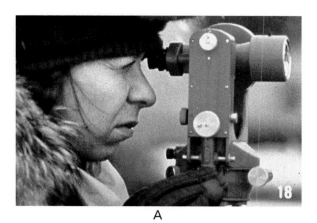

A

B

Fig. 9-16. Surveyors are needed wherever structures are being built. A—This woman worked at surveying the route of the Trans Alaska Pipeline. B—Surveyors find building line for a driveway.   (British Petroleum Co. Ltd.)

port a structure. They bore deep holes and take soil samples at different depths. The holes are called test pits.

Two kinds of soil test may be done at the building site. One is the plate-bearing test. For this test, heavy weights are placed on the surface. This test measures how deep the weights sink into the earth. The test also checks how much the weights tilt.

The second is called a density test. Test soil is placed in a cylinder. Then the soil is hit with a hammer a number of times. This measures now much the soil can be compressed (packed).

Designing

Designing is planning activity. It answers the question: how will the structure be built. Suppose you are going to build a tree house. First, you need an idea how to build it. When you know how it will be built you can decide what materials to buy.

It is the same with any designer. He or she starts with ideas on how to build. Many things come under consideration:
1. How will the structure be used?
2. How strong must it be?
3. What kind of materials will be used?
4. How can I make the structure attractive? Will it "fit" into its surroundings?
5. What will it cost?

A designer must be sure the structure is strong. Foundation and frame must be able to hold the loads placed on them. An engineer computes the weight of the structure. He or she also must know the strength of the materials.

The next step in design is to make drawings, Fig. 9-17. They show the shape and size of every part. They also show where every part belongs.

Designing also includes writing up **specifications.** These are descriptions. They tell how the work must be done. They also explain what quality of materials to use, Fig. 9-18.

Contracting/Managing Construction

On large building projects there are many jobs to be done. Materials must be gathered. People must be told what to do. A contractor

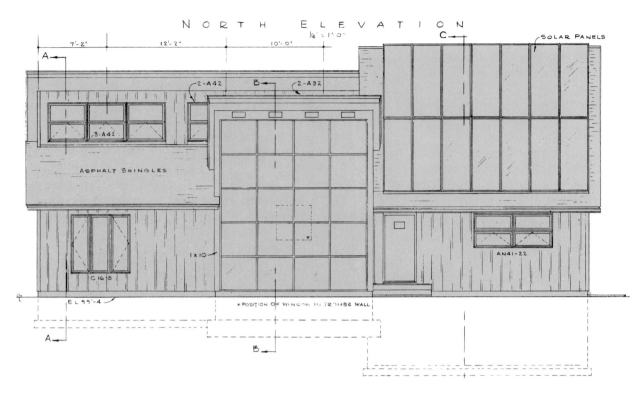

Fig. 9-17. This is one of the working drawings for a structure. It tells the builder the shape and size. (L.F. Garlinghouse)

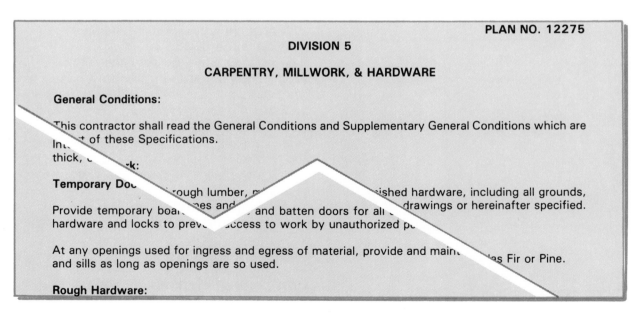

Fig. 9-18. Partial page of a specification for a structure. There may be many pages. Its purpose is to explain what materials to use and the quality of work needed.

is a builder who manages materials and workers. Contractors get work by offering to build for a certain price. They study plans and specifications. Then they enter a **bid.** This is an offer to build the structure for a set price. If an owner accepts a bid, it becomes a contract. A contract is an agreement. The builder agrees to build for so much money. The owner agrees to pay the contractor that price when the structure is built.

Contractors are responsible for hiring workers. They also get materials and equipment. The contractors must make certain that the quality of the work and the materials are good. They must meet the standards set in the specifications.

Hiring subcontractors

Sometimes contractors will pass along part of the work to other companies. The contractor will sign a contract with these companies. For example, one firm will install the electrical system. Another will do all the painting. The subcontractor works for the contractor. The contractor must approve of their work.

CONSTRUCTING

The contractor is just about ready to start building. The owner has purchased the land. The architect has finished the drawings. Work crews have been hired. But wait! Several jobs must be done first.

Getting a permit

Communities want to control what is built. Factories shouldn't go up in a neighborhood of homes. Officials of a city or county want to look at the plans. They want to know that the building is well designed. It must meet local codes (building rules). Then the plans are approved. An official issues a building permit. It is a document which tells the contractor he or she can start work. The permit must be displayed on the building site. It must be in plain view, Fig. 9-19.

Now the contractor and his workers can start to build. Building includes many tasks. Look at Fig. 9-20.

Fig. 9-19. A building permit, left, allows the builder to start work. It means the city has approved the plans. It must be displayed on site until the structure is completed.

Preparing to build

The building site must be cleared. There may be old buildings. These must be torn down. Often they are placed on trucks (sets of wheels) and moved. See Fig. 9-21. Brush and trees are cut. The ground may need to be leveled. Sometimes temporary buildings and service roads are built.

Structures must be located certain distances from the property of others. Work crews must survey. They measure off distances from boundary lines. Measurements may be taken off survey stakes. More stakes are driven to show where to place foundations.

Doing earthwork

The site may be hilly. It must be leveled. If the soil is too loose the structure will settle. The ground must be packed down. Before foundations can be built, large holes and trenches (ditches) need to be dug.

Building foundations

A foundation connects a structure to the earth. It supports the structure so it does not sink into the ground.

A foundation can be constructed of packed earth, gravel, wood, concrete, or steel. Wood

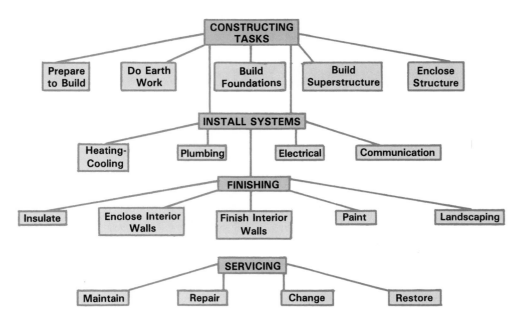

Fig. 9-20. There are many steps in building a structure. They must be performed in the order shown.

Fig. 9-21. Sometimes old buildings must be moved or torn down. This old house will be saved and put on another lot.

or steel poles, called pilings, may be sunk into the ground. They support the structure when the soil is not firm. Fig. 9-22 shows sketches of three types of building foundations.

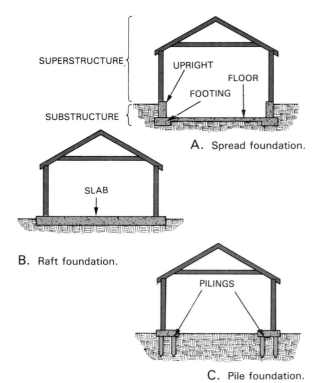

A. Spread foundation.

B. Raft foundation.

C. Pile foundation.

Fig. 9-22. Kinds of substructures (foundations) and superstructures for houses.

Building superstructures

A superstructure is the part of the structure built on top of the foundation. It can be made of steel, concrete, wood, or other material.

Types of superstructures

Look at Figs. 9-22 and 9-23. They show you the superstructure for a house. This type is known as a framed superstructure. Other types are: mass, solid, and air supported. They will be discussed later. Now, let us see how a house is constructed. It is one of the simpler structures to build.

Fig. 9-23. Wood framing used to construct the superstructure of a home. This view shows the interior walls over a plywood subfloor.

BUILDING A HOUSE

A house is a good example of a construction project. It includes all the building steps used in larger projects.

Excavating

Excavating prepares the ground for the foundation. Holes or trenches (ditches) are dug. The foundation must be deep in the ground. Otherwise, freezing and thawing of the earth will damage it.

Excavating can be done by hand, using a spade. This is fine for small construction proj-ects. On larger building projects, a machine is faster. A *back hoe,* Fig. 9-24A, is good for dig-ging wide trenches. It has a shovel attached to a long arm which is controlled by an operator. Fig. 9-24B shows a *trencher.* It scoops out a narrow trench with small buckets attached to a conveyor.

A

B

Fig. 9-24. A—Back hoe digs holes and wide trenches for foundations. B—Trencher is used to excavate (remove earth) for laying of pipe and utility lines.

Building the Foundation

Earlier you learned about types of building foundations. Look at Fig. 9-22 again. Sometimes spread foundations are built of concrete block. But one of the most popular materials for foundations is concrete. It is always used for raft and pile foundations.

Placing concrete

Concrete is a mixture of cement, gravel, and water. Freshly mixed concrete will not hold its shape. It must be poured into some kind of container until it sets. Setting occurs when the water evaporates. Also, a chemical reaction helps *cure* (harden) the mixture. Then the container, called a *form,* can be removed.

Building forms

A form is like a box. But it has no top or bottom, just sides. Sometimes concrete is poured into a trench. Then the earth sides shape and hold the concrete.

The form is designed so it can easily be removed without damage to the concrete. A simple form of 2 in. lumber is used for the footing. Then, more complicated forms are erected for the upright section of the foundation. Fig. 9-25 shows forms being built.

Forms must be carefully leveled. Otherwise the building will not be level or *plumb* (vertical).

Fig. 9-25. Form builders set up forms. Fresh concrete will be poured into them.

Delivering concrete

Concrete can be mixed on the building site. Usually, trucks deliver it premixed. Long spouts, attached to the truck, *convey* (carry) concrete into the forms, Fig. 9-26.

A

B

Fig. 9-26. A—Truck delivers concrete to building site. This slab is being poured for a raft foundation. Note how form holds wet concrete. B—Concrete being dumped into a form for a pillar. Note reinforcing at left.

Sometimes carpenters will build wood forms at the job site. They lay out and cut the lumber. Then they nail them together. Fig. 9-27 shows a section of a form for a foundation wall. Note how it is braced to hold the heavy concrete.

Sometimes foundations, like the one in Fig. 9-28, are built with concrete blocks. No forms are needed. *Mortar* (like cement) binds the blocks together.

Building the Frame

The superstructure (frame) of the house is built on top of the foundation. It has three main parts:
1. Floor.
2. Wall.
3. Roof.

The superstructure must be securely fastened to the foundation. Bolts or straps are embedded in the top of the foundation wall. Wood pieces are fastened to the bolts or straps. These parts make up the **sill plate.** See Fig. 9-29.

Floor framing

The floor frame is fastened on top of the sill plate. The joists stretch from one wall to the

Fig. 9-28. This house foundation was built of concrete blocks. Steel beam is called a girder. It will help support framing members for the floor.

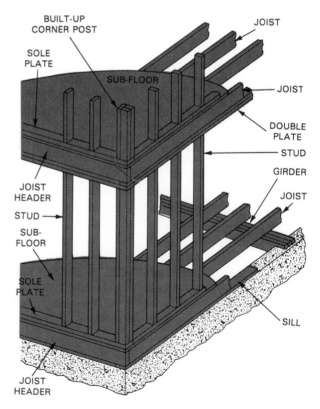

Fig. 9-29. Section of a house frame. Note the names of its parts. Frame pieces are made of 2 in. lumber.

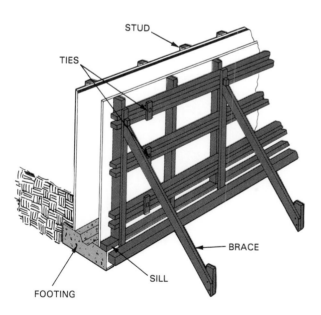

Fig. 9-27. Section of a form for foundation wall. Each part has a name.

other. They carry the weight of flooring materials that are placed on top of them. The joists are held on edge, being nailed to headers at either end.

Sometimes the foundation walls are far apart. The joists cannot span them. Then an extra support is placed between the two walls. This support is called a **girder.** Refer once more to Fig. 9-29. A girder can be either a steel or a wood beam.

Wood girders are usually made by nailing several planks together. The ends of the girder sit on the foundation wall. Posts are also used to help support the weight.

Wall framing

In most houses walls are made of wood framing. Wall frames have many parts. There are plates, studs, and headers. The frames are light but strong, Fig. 9-31.

Plates are the horizontal parts at the top and bottom of the wall. They hold the **studs** which are the vertical parts. Studs are spaced 16 or 24 in. apart. The spaces between the studs are later filled with insulation.

Roof framing

Wood-framed buildings may have flat or pitched roofs. A pitched roof is lower at the walls. It rises several feet to the center of the house. The high part of the roof is called the **ridge.** The low end is called the **eave.**

The roof framing members are called **rafters.** They are generally spaced 16 to 24 in. apart. See Fig. 9-30.

A

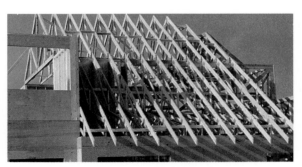
B

Fig. 9-31. A—A closed in house. Doors and windows have been installed. B—Door being installed. It is preassembled. It is slipped into place and fastened to the building frame. (Stanley Door System)

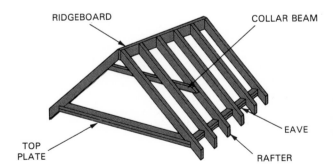

RIDGEBOARD   COLLAR BEAM

TOP PLATE

EAVE

RAFTER

Fig. 9-30. A roof frame and its parts. Note collar beam which keeps rafters from pushing walls outward. Rafters at right are called "truss" rafters.

At the ridge, the rafters are fastened to a **ridge board.** This keeps the rafters spaced properly. **Collar beams** are fastened to the rafters below the ridge board. They brace the rafters.

Closing in the Building _____

The building frame just described will support all the materials used to close it in. Enclosing:
1. Keeps out rain, snow, heat, and cold.
2. Protects the inside of the building and the people who use it.
3. Gives privacy.
4. Makes the building attractive.

Builders will perform many tasks to close in the structure. They will cover the frame with plywood, flakeboard, oriented strand board, or insulating board. These materials are made up in 4 by 8 ft. sheets. Thickness varies, but 1/2 and 3/4 in. panels are common. Applied to outside walls, such materials are generally called sheathing. Sheathing used on roofs is known as decking. Sheathing not only closes in the structure, it makes the frame stiffer.

After sheathing, carpenters will install outside doors and windows. These generally come assembled. But carpenters must fit them into the openings of the sheathed frame, Fig. 9-31.

Attaching exterior finishing materials ___

Sheathing only closes up the openings in a building frame. Attractive weatherproof materials cover the sheathing, Fig. 9-32. The material could be brick, stone, wood siding, aluminum, vinyl (plastic), wood panels, or shingles.

Applying roofing materials _____

Roofs are more likely to leak than walls. Special waterproof materials must be used. Shingles are normally used on all pitched roofs. On flat roof surfaces a roof covering must be built up from layer upon layer of materials.

Shingled roofs _____

A shingled roof is installed by attaching many small pieces of material that are waterproof. They also withstand many kinds of weather conditions. As each piece is installed, it overlaps the one beneath it. It looks somewhat like the scales on a fish. See Fig. 9-33. Shingles are made of many different materials; wood, slate, ceramic, aluminum, fiberglass, or asphalt.

Built-up roofs _____

Flat roof surfaces cannot use shingles. Water would run under them and cause leaks. A built-

A

B

C

D

Fig. 9-32. Siding is made of many different materials. A—Hardboard siding. B—Bricklayers install brick exterior wall. C—Carpenter cuts and shapes aluminum siding. D—Painter covers wood siding with protective, decorative coat of paint. (Rohm & Haas)

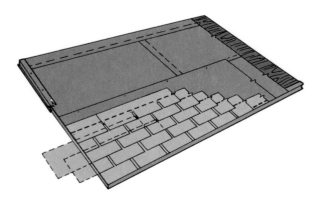

Fig. 9-33. Shingles are nailed to the roof. See how they overlap. This makes the roof waterproof. (Asphalt Roofing Manufacturers Assn.)

up roof must be installed. It is made with layers of building felt, an asphalt (tarlike) material called bitumen, and gravel.

First, roofers will lay down a thick layer of hot liquid bitumen. Then they will roll out building felt over the bitumen. This process will be repeated until there are three, four, or more layers. Finally, gravel will be spread over the top layer of bitumen. The bitumen, liquid when hot, hardens to a tough but flexible material. It glues the felt and gravel together. It provides a waterproof fire-resistant coating, Fig. 9-34.

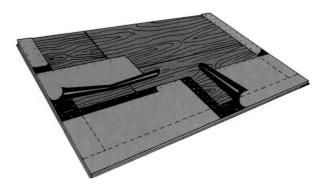

Fig. 9-34. Flat or nearly flat roofs will leak if shingles are used as a covering. Overlapping sheets of tough waterproof material are ''tarred'' with asphalt. Sometimes, on flat roofs, gravel is poured over the top.
(Asphalt Roofing Manufacturers Assn.)

A plastic fabric is sometimes used on a flat roof. It replaces the asphalt materials and the gravel. Called a membrane, it is tough, durable, and waterproof without the use of bitumen.

Installing Utilities

Utilities are services that come into a building. They are supplied from pipes and cables, Fig. 9-35. Most buildings will have electrical wiring. They will also have a plumbing system to bring in water and carry away wastes. An air conditioning system supplies cooled air. Sometimes piping is needed to bring in gas for heating. These systems are hooked up to the outside utility lines. These lines are usually buried in the street. They are supplied by the community.

In a building, these systems are usually called the mechanical systems. These systems must be installed before the inside of the house is finished.

Roughing in mechanical systems

Installing mechanical systems is called roughing in. Usually you cannot see the systems. They are placed inside the walls and under floors. If you go into the basement you can sometimes see the pipes, ducts, and electrical wiring.

Roughing in is usually carried out in a special order. Ductwork or pipes carry heat from a furnace or boiler. They go to each room of the house. They are put in first because they are very large. They are fitted between floor joists and wall studs. Fig. 9-35 shows a furnace and simple duct system.

Plumbing systems are installed next. This system of pipes carries liquids and gases. Plumbing in a building is needed to:
1. Provide people in the building with fresh, pure water.
2. Remove waste water.
3. Carry fuel gases to furnaces, water heaters, and stoves.

Like the ductwork, plumbing is installed in the open spaces between the wall studs and the floor joists, Fig. 9-36. Sometimes holes and notches must be cut to fit the pipes.

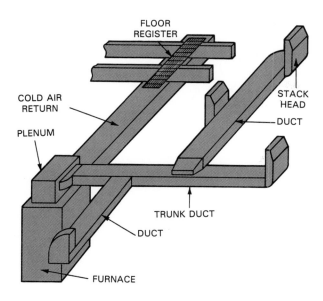

Fig. 9-35. Ducts carry conditioned air from a furnace or air conditioner to all rooms of a building.

Fig. 9-36. Plumbing system pipes supply fresh pure water. Other pipes carry waste water.

Pipes that supply fresh water or gas can be quite small. Larger pipes are needed to carry away wastes from bathrooms and kitchens.

Electrical systems distribute electrical power. They also are used for communication. Electric power runs appliances in a home. It also is used for lighting. Telephones and intercom systems use electrical impulses to carry speech.

Electricity can be dangerous. Current flows through wires which are insulated. The insulation keeps the electric current from damaging the building. It also protects people from shock. It must be installed according to special safety rules. These rules are called the National Electrical Code.

Electrical wiring is small. It is installed last. Sometimes the wires are carried inside a lightweight pipe called **conduit.** The wires can also be embedded in a cable with a plastic covering. This type is called nonmetallic cable. A third kind of wiring system is called armored cable or "BX." Like ductwork and plumbing, electrical wiring is placed between studs and joists. See Fig. 9-37.

Insulating

Outside walls and ceilings of the house must be insulated. Insulation is a material that resists heat passage through it. Insulation keeps heat inside during cold weather. In hot weather it keeps the heat outside. It is very light and soft because it is full of air pockets. It is made in several forms. Loose insulation can be blown into ceilings. It also is made in long strips called batts. These are placed between studs and joists. See Fig. 9-38.

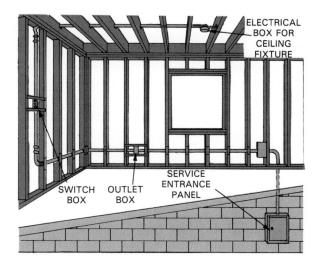

Fig. 9-37. Electrical conduit, or cables, are placed in walls and ceilings. Wiring is connected inside small metal or plastic electrical boxes.

Fig. 9-38. Insulation is made from materials that do not conduct heat. It is placed in walls and ceilings. It prevents heat from passing in or out of a building.

Closing and Finishing Interiors

Mechanical systems have been installed. Insulation is in. Now, workers enclose the interior (inside). First walls and ceilings must be covered. Then finishing work starts.

Closing in

At one time walls and ceilings were covered with plaster. It was applied over thin, narrow slats called lath. Today, drywall is used in most buildings. It is manufactured by putting a layer of gypsum (like chalk) between two layers of heavy cardboard. Drywallers attach large sheets of drywall with glue or nails. Seams and nail heads are concealed. A special tape and filler are used.

Finishing work

Finishing includes jobs that make the interior of the building attractive. These tasks, too, follow a certain order:
1. Painting and decorating.
2. Installing finish flooring.
3. Installing trim.
4. Installing fixtures and accessories.
5. Cleaning up.

Painting and decorating is done to beautify the interior. Plaint protects wood and drywall surfaces. Cleaning is also easier. But wallpaper, paneling, and tile are other choices for wall coverings. See Fig. 9-39. Paneling has become a popular and durable covering. Paneling comes from the factory already stained and finish coated.

Fig. 9-39. A carpenter installs paneling to cover concrete block. Drywall can be protected and beautified with paint. (Wagner Spray Tech Corp.)

Putting down finish flooring

Finish flooring is designed to wear well and look attractive. Builders usually install it after the painting and decorating is done. It avoids damage to and soiling of the floor material.

Many different materials make good finish flooring. Wood, carpeting, linoleum, and tile are among the most used. They are fastened to the subflooring or underlayment with special nails or adhesives (glues). See Fig. 9-40.

Fig. 9-40. Installing flooring materials. These may be sheet materials, or as shown, individual tiles. Carpeting is another popular material for covering floors. (Armstrong World Industries, Inc.)

Trimming out and installing accessories

Trim is the decorative wood or plastic strips that cover joints. Joints are the cracks where floors, walls, and ceilings meet. There are also cracks where window and door frames meet walls. Carpenters must make accurate joints where trim pieces meet. Next, doors are carefully hung on their hinges. Edges are dressed (planed) so they fit openings.

Finally, hardware and accessories are installed. Hardware means door knobs, latches, catches, towel bars, and brackets. Accessories can be major or smaller items. Major items include kitchen cabinets, bathroom cabinets, or built-in units. Other accessories installed at this time are counter tops and closet shelving. Plumbing faucets, electrical fixtures are among the final items installed.

Finishing the Site

Construction is not complete until the building site is finished. During construction the site becomes cluttered. Scraps of building materials are here and there. There may be piles of dirt. Several things must still be done.
1. The site is cleared.
2. Ground may need leveling or grading.
3. Concrete walks and drives are poured.
4. Landscaping is added.

Clearing the site

Some clearing of the site may need to be done before any other finishing steps are taken. Some of the dirt may not be needed. It must be hauled away. Rocks, trash, and scraps of building materials must be removed. Temporary buildings and fences are taken down.

Leveling and grading

Earth may have to be moved. Holes may have been dug for foundations. Some of the earth is pushed back to fill in around the foundation. This is called backfilling. The earth is shaped around the structure. Soil may be moved from one spot and placed in another so it is more pleasing.

Installing walks and drives

Walks and drives give users access to a building. Drives must have a heavy base of gravel. The surface is concrete or asphalt. Sometimes no other surface but the gravel is needed. Sidewalks can be constructed of concrete, natural stone, wood, or masonry units.

Landscaping

Landscaping is a way of making the site more attractive. It includes planting trees, shrubs, and flowers. It also often includes planting grass or putting down sod, Fig. 9-41. Ground cover is sometimes used where grass is not wanted. It keeps soil from washing away. It also covers unattractive soil. Usually, ground cover is a low-growing plant. But is can also be wood particles, bark, or stone.

Fig. 9-41. Landscaping is usually done by specialists. They grade and fertilize the earth, plant trees and shrubs and lay down sod.

Servicing Structures _____

From time to time structures must be maintained. When a roof begins to leak it must be repaired, Fig. 9-42. Repainting also helps to preserve many structures.

OTHER CONSTRUCTION

As you have learned from this chapter, buildings are only one of the many types of structures. Fig. 9-43 is a simple design of foundations and superstructures for roads and dams. These are called mass foundations and mass superstructures. Fig. 9-44 shows a road under construction. Fig. 9-45 is part of the superstructure for a new bridge. Fig. 9-46 is a type of framed superstructure used for large buildings with many stories. Fig. 9-47 is a special building. Its roof is supported by air!

Constructing in space _____

What about constructing things in space? For more than 10 years science and technology ex-

perts have been talking about it. They have been designing structures that might be used.

Eventually, there will be much work to do in space. There will have to be living quarters. Work areas will be needed to service space vehicles such as satellites, Fig. 9-48.

Parts of the structures can be built on earth. Then they will be sent into orbit. Parts will be assembled in space by workers in space suits. They will need special tools and materials. These will have to be invented.

This type of construction is a part of the future. Perhaps you will be part of a team that will design and build in space.

SUMMING UP

Construction is building or putting a structure together at the spot where it will be used. Our ancestors knew little or nothing about putting up buildings. They lived in caves or built crude shelters that did not last. In time, people began to settle in one place and erected per-

Fig. 9-42. Left. These workers are installing a new roof on an office building. Right. Paint preserves this smokestack and carries the company logo (name).

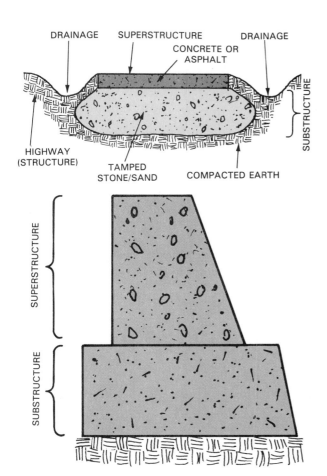

Fig. 9-43. Other types of foundation and superstructure. These are known as "mass" structures.

Fig. 9-44. Earth is moved and compacted to make the foundation for a road. Then the asphalt superstructure is laid down. (Natchez Trace Parkway, National Park System)

Construction Systems 185

Fig. 9-45. Top. The superstructure of this bridge is built of reinforced concrete pillar. It will stand for years. Other bridge parts are made of steel. Bottom. Pouring concrete for a marina super-structure.   (Prestressed Concrete Institute)

Fig. 9-46. Two types of framed superstruc-tures. Top. An office building in St. John, New Brunswick, is getting a reinforced concrete superstructure. Bottom. Steel pillars and beams will support the roof and walls of this warehouse.

Fig. 9-47. Did you ever see a building built like a balloon? This is a sports arena in Indianapolis. The walls are of solid construction but the roof is supported by nothing more than air pressure.

Fig. 9-48. An artist's idea of what it will be like building in space. In the space shuttle's cargo bay is a beam builder. It automatically builds triangular girders from very light metal plate. (NASA)

manent buildings. Important ancient builders were the Greeks and the Romans. They were famous for their beautiful buildings.

Structures of today are built for several purposes. Residences and commercial buildings shelter people where they live and work. Civil structures help move people and materials or give some other benefit to the public.

Resources for construction are people, material, and knowledge of construction methods.

Construction processes include acquiring land, designing the structures, excavating, putting in foundations, and building the frame or superstructure. It also includes installing the mechanical systems and finishing.

KEY WORDS

These words are used in this chapter. Do you know their meaning?

Bid, Boundaries, Bridging, Building permit, Chain, Civil structure, Collar beam, Conduit, Contract, Contractor, Eaves, Girder, Mechanical systems, Negotiate, Rafters, Ridge board, Sill plate, Specifications, Studs, Superstructure, Surveying, Utilities.

TEST YOUR KNOWLEDGE
———————— Chapter 9 ————————

Do not write in this text. Place answers to test questions on a separate sheet.

1. Which of the following statements best tells what construction is?
   a. Building a road, dam, bridge, pipeline, monument, or building.
   b. Building a structure at the spot where it is going to be used.
   c. Building all the things we use to protect us and make our lives more comfortable.
2. Constructed projects are always permanent. They cannot be moved from the site. True or False?
3. Describe a common temporary shelter built by primitive people who moved often.

MATCHING TEST. Match the term with the right description:

4. ____ Structures meant to provide protection for human activities.
5. ____ Provide protection for living activites.
6. ____ Provide a place for carrying on business or education.
7. ____ Provide for manufacturing activities.
8. ____ Provide for movement of people and material and support other activities necessary for living, working; provide beauty and culture.

A. Commercial, public, and institutional buildings.
B. Shelters.
C. Residential buildings.
D. Civil structures.
E. Industrial building.

9. Tell to which category each of the following structures belong (residential, commercial or public, industrial, and civil):
   a. _____ Power lines strung on towers from a power plant.
   b. _____ A trailer home in a mobile home park.
   c. _____ A building where people work to produce electric toasters.
   d. _____ An expressway between two cities.
   e. _____ A college dormitory.
10. What are resources?
11. All the information needed to build a structure is contained in the (select best answer):
    a. Cost estimate.
    b. Plans and specifications.
    c. Plans, specifications, and construction knowledge of craftspeople.
    d. Building codes.
12. Name five kinds of building materials.
13. When a buyer and seller get together to agree on the price of a piece of land it is called _____.
14. Arrange the following steps in their proper order for preparing a building site (use letters of the alphabet):
    _____ Clear the site of unwanted trees and brush.
    _____ Excavate.
    _____ Dig test pits.
    _____ Get a building permit.
    _____ Tear down an old building.
15. In building a foundation, a wide, flat base, called a _____ is first constructed; then vertical supports called _____ are built.
16. When large masses of material are piled up for a superstructure, it is called a _____ superstructure.
17. It is not possible to use air to support superstructures. True or False?
18. Until it can harden, concrete must be poured into a _____ to keep its required shape.
19. Joists and headers are parts of:
    a. Wall framing.
    b. Roof framing.
    c. A form for concrete.
    d. The floor frame.
    e. Are not used in construction.
20. Inside a building, the heating ducts, electrical wiring, and plumbing pipes are known as the _____ _____.

ACTIVITIES

1. On your way to school look for construction projects. Make a list of them. Try to separate them into one of the following groups: residential building, commercial building, civil construction, or industrial building.
2. Select a structure you pass every day and write a paragraph that explains its purpose.
3. Try to list the reasons your school is located where it is.
4. Write a short report on how your home is heated. Ask one of your parents or an older member of your family to help.

APPLYING YOUR KNOWLEDGE

Introduction

There are structures everywhere you look. There are houses, schools, hospitals, stores, and factories. Around them are streets and roads, parking lots, and power lines. Underground there are storm sewers, water mains, and gas lines. Elsewhere there are airports, railroad lines, dams, and bridges. These are all part of our "constructed" world. They are the results of people using construction technology.

In this activity, teams in your class are going to work together to construct a structure. They will use construction technology to build a model of a storage shed.

Equipment and Supplies

Scale lumber and building materials:
  1/4" x 5/8" x 16" pine (scale 2 x 4 x 8)
  1/4" x 1" x 16" pine (scale 2 x 4 x 8)
  1/8" x 5/8" x 16" pine (scale 1 x 4 x 8)
  8" x 16" 6-ply posterboard (scale 1/4" x 4' x 8' plywood)
  8" x 16" matboard (scale 1/2" x 4' x 8' plywood)
Rules
Squares
Miter boxes
Back and coping saws
Utility knives
Hammers
5/8" x 18 brads
Adhesives

Procedure

Your teacher will divide the class into groups to construct the various parts of the shed. These groups and their tasks will be:
  Group 1 — Front wall.
  Group 2 — Left side wall.
  Group 3 — Right side wall.
  Group 4 — Rear wall.
  Group 5 — Floor.
  Group 6 — Rafters.
Each group should:
1. Carefully study the plans provided. These include:
   Fig. 9A — Pictorial view.
   Fig. 9B — Floor plan.
   Fig. 9C — Floor joist plan.
   Fig. 9D — Front elevation.
   Fig. 9E — Left side elevation.
   Fig. 9F — Right side elevation.
   Fig. 9G — Rear elevation.
2. Prepare a material list for the section your group is to build. You may need to study several of the drawings to determine the sizes of all materials. List each part, the number needed, the size shown on the drawings, and the scaled size of material you will need. For example:

No. | Part | Actual size | Scale size
8 | studs | 2 x 4 x 6' | 1/4" x 5/8" x 12"

   NOTE: The actual length is divided by 6 to determine its scale length.

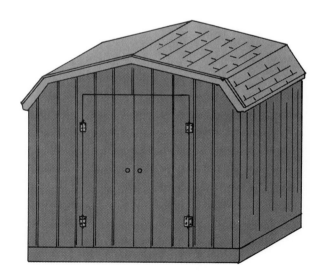

Fig. 9A. Pictorial drawing of the storage shed.

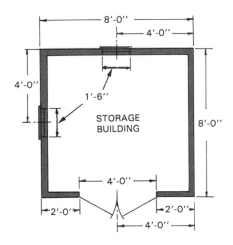

Fig. 9B. Floor plan.

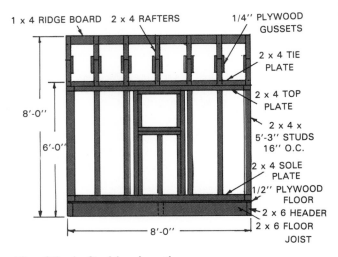

Fig. 9E. Left side elevation.

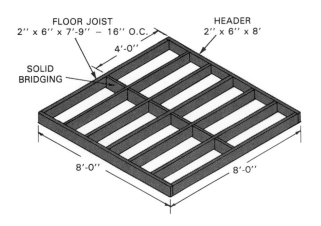

Fig. 9C. Floor joist plan.

Fig. 9F. Right side elevation.

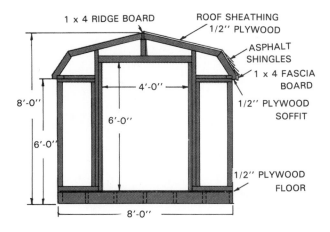

Fig. 9D. Front elevation.

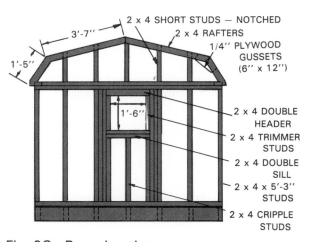

Fig. 9G. Rear elevation.

190   Understanding Technology

3. Get the materials your group needs to make the shed section.
4. Cut all materials to their correct size.

SAFETY NOTE: Be careful when using hand tools with cutting edges. Never touch the cutting edge with your hand. Cut away from any part of the body. Carry sharp-edged and pointed tools turned downward and away from the body. Never carry sharp tools in your pockets. Store tools not in use. Always check with your instructor for safety instructions with any tool.

5. Have your teacher check the material.
6. Assemble the assigned shed section according to the drawings. Be careful to observe the following precautions:

GROUP 1 — Front wall, Fig. 9H.
   a. The tie plate is 5/8 in. shorter than the top plate on both sides of the door. This lets the side section tie plate overlap and connect the sides to the front.
   b. There are cripple (short) studs under the door header.

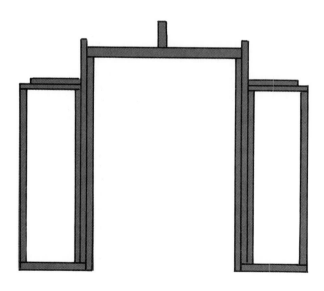

Fig. 9H. Front wall assembly drawing.

GROUP 2 — Left side wall, Fig. 9I.
   a. There are cripple studs between the window header and sill.

   b. The drawing appears to call for three studs on each side of the window. Actually there are two studs separated by a space.
   c. The tie plate extends 5/8 in. beyond the end of the top plate.

Fig. 9I. Left side wall assembly drawing.

GROUP 3 — Right side wall, Fig. 9J.
   a. This wall is different from that of the left side wall.
   b. The tie plate extends 5/8 in. beyond the end of the top plate.

GROUP 4 — Rear wall, Fig. 9K.
   a. The tie plate is 5/8 in. shorter than the top plate on both sides of the door. This allows the side section tie plate to overlap and connect the sides to the front.
   b. There are cripple studs between the window header and sill.
   c. The drawing appears to call for three studs on each side of the window. Actually there are two studs separated by a space.

GROUP 5 — Floor, Fig. 9C.
   a. The floor joists are fabricated first.
   b. The floor joist assembly is then covered with 1/2 in. plywood (matboard).

GROUP 6 — Rafters, Fig. 9L.
   a. Four sets of rafters have 1/4 in. plywood (posterboard) gussets on on-

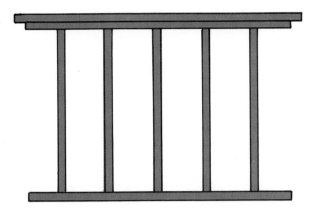

Fig. 9J. Right side wall assembly drawing.

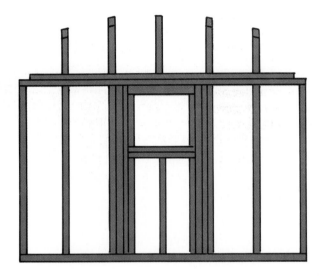

Fig. 9K. Rear wall assembly drawing.

ly one side. Be sure that you make two left side and two right side rafters.

  b. All other rafters have gussets on both sides of the joist.

7. Have your teacher check your constructed sections or parts.
8. Assemble the shed:
  a. Secure the rear wall to the floor.
  b. Secure right and left walls to floor.
  c. Secure the front wall to the floor.
  d. Secure the rafters and ridge board in proper position.
9. Finish the shed by applying:
  a. Siding.
  b. Roof sheathing.
  c. Shingles (abrasive paper).
10. Fabricate and install a door. (optional)
11. Install a window — plastic square. (optional)
12. Paint the shed. (optional)

Fig. 9L. Rafter assembly drawing. See Fig. 9G for dimensions.

Chapter 10
Communication Systems

The information given in this chapter will help you to:

- ☐ Describe several ways people communicate.
- ☐ Define communication technology.
- ☐ List and describe the types of communication technologies.
- ☐ Describe the types of graphic communication systems.
- ☐ Describe the types of electronic (wave) communication systems.
- ☐ List and describe the four types of communication.
- ☐ Diagram and explain the communication process.
- ☐ List and describe the four-step process for producing a message.

Humans communicate constantly. They talk to one another. Persons read what other humans write. They listen to voices from radios. They view and listen to television programs. In fact, humans cannot avoid communication. Try it. Can you see anything? Do you hear anything? Can you smell anything? Are you touching anything? Are you tasting anything? If you are, you are receiving information. And that is **communication.** Anytime people send or receive information they are communicating. Think about how many times you have received or sent a communications message already today.

HOW WE COMMUNICATE

People must process this information before they can communicate. You may know something. The knowledge is in your mind. No one can see it. But you can tell a friend what you know, Fig. 10-1. You could draw a pic-

Fig. 10-1. People can communicate with pictures, symbols, writing, or speaking.

ture which communicates the information. The picture shows your friend what you know.

You could also use a symbol. You could show your friend a sign that represents the information. The highway signs, as shown in Fig. 10-2, are symbols. They provide information to people who drive cars.

Fig. 10-2. Highway signs are also used to communicate.

Also, you can pass along information through speech. You could make sounds that have meaning. As long as two people know what each sound means they can communicate. Yet, not everyone understands all spoken words. You may say, "I am 12 years old," to a group of Chinese people. They will understand you only if they speak English.

Finally, you may write your information. You may use a special set of symbols. They are called the alphabet. By using the letters of the alphabet, you can communicate information. Again, the arrangement of letters will change from one language to another. To us a four

legged animal that barks is a "dog." In Spain it is called a "perro." In Germany it is a "hund."

So far we have discussed people directly communicating with other people. This is face-to-face communication. They show pictures and symbols or they use formal language. These are common types of communication. But they use little technology. There are no technical means used. Machines or equipment are not involved in the communication. A communication system is not present.

COMMUNICATION TECHNOLOGY

Communication technology uses equipment and systems to send and receive information, Fig. 10-3. This technology communicates information using **graphic** and **wave** systems, Fig. 10-4. (Graphic comes from a word meaning to draw or write as on paper. "Wave" refers to radio waves, a kind of energy.)

Fig. 10-3. Communication technology uses equipment and systems. (Gannett Co.)

Graphic Communication

The information may be drawings, pictures, graphs, photographs, or words on flat surfaces, Fig. 10-5. This type of message is called **graphic**

GRAPHIC	ELECTRONIC

Fig. 10-4. The two major communication systems are graphic and electronic (wave) systems.

Fig. 10-5. A newspaper is an example of a graphic communication media.   (Gannett Co.)

communication. The **media** used is two dimensional. It has length and width. Media is a name for matter which carries the message. In graphic communication sheets of paper or film are the carriers. But in wave communication, radio waves are the media. Graphic communication systems produce printed graphic and photographic media. (They communicate through drawings, printed words, and pictures.) The systems also use photographic prints and transparencies (slides and motion pictures). These messages are visible as they move from the sender to the receiver.

Wave (Electronic) Communication

Wave communication systems depend on an energy source called **electromagnetic radiation.** This is energy that moves through space in waves. Heat and light from the sun are two examples of electromagnetic radiation. Look at Fig. 10-6.

All electromagnetic waves travel at the speed of light — 186,000 miles per second.

Sound waves are much slower than electromagnetic waves. They travel at different speeds in different media. In air they travel at about 750 miles per hour at sea level. They travel about four times faster in water. Sound travels fastest in hardened steel. There it travels at 16,000 feet per second or about 10,000 miles per hour.

People can use light, sound, or electrical waves to send information. The information is coded at the source. It is then transmitted

Communication Systems   195

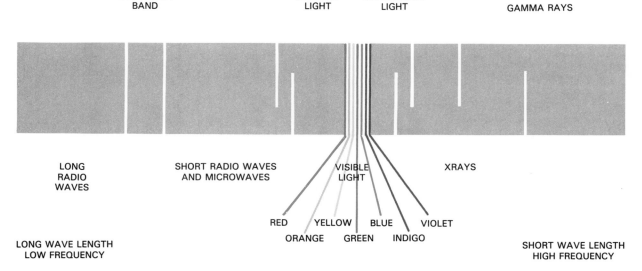

BROADCAST BAND   INFRARED LIGHT   ULTRAVIOLET LIGHT   GAMMA RAYS

LONG RADIO WAVES   SHORT RADIO WAVES AND MICROWAVES   VISIBLE LIGHT   XRAYS

RED   YELLOW   BLUE   VIOLET
ORANGE   GREEN   INDIGO

LONG WAVE LENGTH LOW FREQUENCY   SHORT WAVE LENGTH HIGH FREQUENCY

Fig. 10-6. The electromagnetic spectrum.

(sent) to the receiver. There the code must be changed back to information. This type of communication, shown in Fig. 10-7, is often called electronic communication. These systems often use electronic equipment to code and transmit the information.

For example, suppose that your favorite sport figure is interviewed, Fig. 10-8. He or she speaks into a microphone. Speech is changed into pulses of electromagnetic energy. This energy is transmitted from the radio station's broadcast tower. Your radio's antenna picks up some of these pulses of energy. The energy is changed by various stages of the radio. Finally the radio speaker changes electrical pulses back into *audible* (can be heard) sound. You hear the voice. But, between the microphone and the speaker no one can hear or see the message.

Likewise, many telephone messages are changed into pulses of light. These are carried along glass fibers to their destination. The pulses of light are changed back to sound waves. This process is called fiber optic communication.

Sonar uses sound waves to locate submarines, fish, and other objects under water. Sound waves are transmitted from a ship. They strike another vessel and bounce back to the ship. The time it takes the sound to travel is recorded. This allows an operator to determine the position and distance of the second vessel. Sound is also used in the search for oil and mineral deposits. Geologists use this in a process called seismic prospecting.

Graphic vs Electronic Communications

Electronic communications have many advantages over graphic communication. A book can hold only so much information. Readable type can be only so small. Also, a book can be only so thick and still be useful. But wave communications carry vast amounts of information in almost no space. A space satellite communicates a library of information in a matter of seconds. Fiber optic systems are said to carry encyclopedias of information per second.

But you can't take wave media everywhere with you. There isn't always a telephone or television set next to you. A signal from a radio station can only reach so far. But books and magazines are portable. They can also be available at any time or place. They capture ideas and visual images. On the other hand, they cannot provide sound like music.

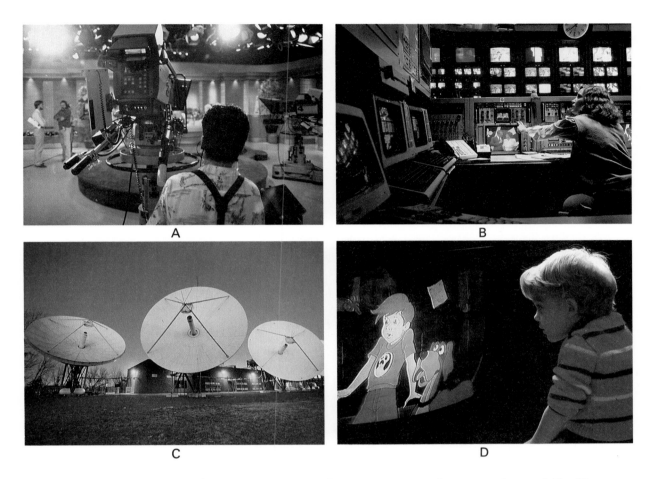

Fig. 10-7. A typical communication system. A—Cameras capture pictures and sound. B—These signals are processed by complex equipment. C—They are then sent through the air. D—The viewer gets the message on a receiver (TV set). (Westinghouse Electric Corp.)

Fig. 10-8. Radio brings voices from thousands of miles into our homes. (Gannett Co.)

Therefore, we need different types of communication systems for different uses. Wave communication is quick and cheap. Current events reach us almost as they happen. Print communication media are slower but can be used almost everywhere. They can be easily selected, used, and stored. And, they will last a lifetime.

TYPES OF COMMUNICATION SYSTEMS

Communication technology is used for four distinct types of communication. These, as

shown in Fig. 10-9, are:
1. People-to-people communication.
2. People-to-machine communication.
3. Machine-to-people communication.
4. Machine-to-machine communication.

Each of these communication systems affects our daily lives. We come into contact with each type as we interact with our environment.

All the systems have similar components, Fig. 10-10. They all have a device to develop and send the message. This device is often called a **transmitter.** All systems must have a channel to carry the message. This channel is called the **carrier.** It can be the airways, a wire, or optical fibers. Finally, each system has a device to gather and process the message. This device is generally called a **receiver.**

People-To-People Communication

Everyone communicates with other people. Humans have developed complex technological

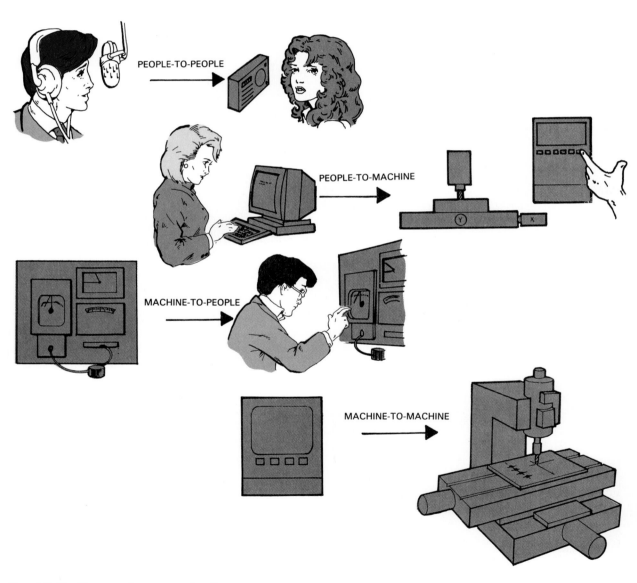

Fig. 10-9. Types of communication.

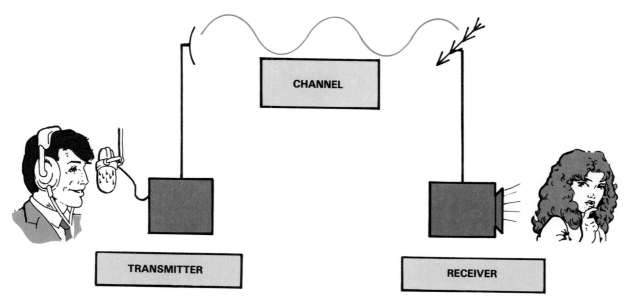

Fig. 10-10. A simple communication system.

systems to improve this type of communication. Machines and devices have been produced to help us communicate better and easier.

People-to-people communication includes five basic systems. These are:
1. Telecommunication systems.
2. Audio and video recording systems.
3. Printing systems.
4. Photographic systems.
5. Drafting systems.

Each of these systems is in wide use today. Each has its place and serves a specific purpose.

Telecommunication systems

Communication systems which exchange information over a distance are called **telecommunication.** There are two major types of telecommunication systems. These are individual and mass communication systems.

Individual telecommunication systems have been designed to let one person communicate to another individual. These were the earliest telecommunication technologies. Two systems of this type were developed in the 1800s.

The first was the work of Samuel Morse in the 1830s. It was the **telegraph.** This system uses an electrical circuit (complete pathway through which electricity flows) to carry the message. (The circuit is the carrier.) An operator uses a special code, Fig. 10-11. A series of dots (short electrical pulses) and dashes (long pulses) represent letters and numbers. The operator produces the pulses by pressing a key (transmitter). At the other end of the circuit a sounder (receiver) is activated. It changes the pulses into clicks which are read as dot or dashes.

The second personal telecommunication systems was the telephone, Fig. 10-12. Alexander Graham Bell invented this device in 1876. It uses an electrical circuit to connect the two communicator's phones with a central office (switching area). The handset contains a transmitter which changes sound waves into electrical pulses. It also contains a receiver which changes the electrical pulses back into sound.

Often the two telephones are connected to different central offices. Then the message is transmitted first to the central office on wires. From there the signal moves between the two offices. This can be done on wires, by microwaves (high frequency waves) between towers or a satellite, or by glass fiber cables (fiber optics).

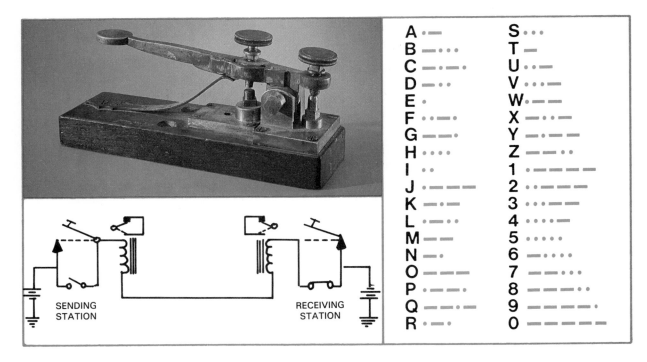

A	·—	S	···
B	—···	T	—
C	—·—·	U	··—
D	—··	V	···—
E	·	W	·——
F	··—·	X	—··—
G	——·	Y	—·——
H	····	Z	——··
I	··	1	·————
J	·———	2	··———
K	—·—	3	···——
L	·—··	4	····—
M	——	5	·····
N	—·	6	—····
O	———	7	——···
P	·——·	8	———··
Q	——·—	9	————·
R	·—·	0	—————

Fig. 10-11. Top left. An early telegraph. Bottom left. Schematic of a closed loop telegraph system. Right. The Morse code.

Fig. 10-12. The telephone is a business and personal communication device. (Ohio Art Co.)

Radio and Television

Two mass telecommunication methods have been designed to reach large groups of people. These are radio and television, Fig. 10-13.

They operate very much alike. They change sound (audio) and/or light (video) waves into high frequency signals. These systems use microphones and cameras to change sound and light into electrical pulses. Transmitters then process these pulses into high frequency waves. These signals are carried to a broadcast tower. There they are *radiated* (projected) over a large area. The atmosphere serves as the carrier for the signal. An antenna captures some of these waves. The receiver separates the desired signal from other waves. The signal is then changed into sound and light by speakers and television picture tubes.

Radio and television broadcast systems use two types of signals. The VHF (very high frequency) bands (groups of signal frequencies) are used by radio and television stations. The UHF (ultra high frequency) bands are used by some television stations.

Audio and video recordings

Audio and **video recordings** are extensions of radio and television communications. They record the same type of information that radio

Fig. 10-13. The control room, left, is the nerve center of a television broadcasting activity (right). (Viacom International, Inc., and Gannet Co.)

and television broadcast. They can be seen as a warehouse for audio and video information.

These communication systems can be traced to Thomas Edison's invention of the phonograph, Fig. 10-14. This 1877 invention allowed people new freedom. No longer did they have to listen to a talk or concert when someone else scheduled it. They could select the time and place they wanted to listen to the music or speech.

Audio and video recordings use one of three technologies:

1. Grooves in discs. This is the recording technology used by standard audio records. Records are produced with wavy grooves on their surface. A *stylus* (needle) vibrates as it follows the grooves on a moving record. This vibration produces a small electrical signal. This signal is amplified (made stronger) and changed into sound waves.

2. Magnetic charges on a tape. This is the technology used for audio and video tapes. Coated plastic ribbons are fed through a recording unit. The unit produces a pattern of magnetic charges in the coating. The playback unit "reads" the charges. They are then changed into sound (audio) and light (video) messages.

3. Digital codes on a disc. This is the technology used to produce compact audio and video discs, Fig. 10-15. Sound and light

Fig. 10-14. This is an early model of the Edison phonograph.

Fig. 10-15. This person is loading a laser video disc into a playback unit. (Rohm and Haas)

waves to be recorded are **digitized** (numbered). Each frequency is assigned a specific nuumber. The number is recorded on the disc on microscopic (very small) pits and flats. The playback unit reads the code with a laser beam. The digital readout is then converted back to very accurate pictures or sound.

Each of these systems has its advantages and disadvantages. Disc recordings are cheap to manufacture. But, they scratch easily and lose quality. Tape recordings are small and unbreakable. Also, new material can be recorded over a used tape. But they lose quality over time and sometimes jam in the player. Compact discs produce high quality sound reproduction-but are fairly expensive.

Communication in space

A **communication satellite** is a broadcasting station in outer space. It can receive a message from one place on earth. Then it relays or "reflects" the message back to earth at another distant place. Look at Fig. 10-16.

The message is carried by a narrow beam of microwaves. The beam has to be carefully aimed. The signals sent from earth must hit the satellite. They can be beamed back to earth at many different anagles and to more than one earth station.

Because of the need to aim the narrow beam, the satellite must always be in the same position above the earth. It must complete one orbit every 24 hours. For this reason its orbit is called geosynchronous (at the same time as the earth). If it were to travel slower or faster the microwave link (contact) would be broken.

Communication satellites are placed about 22,300 miles above the earth. There they can transmit to over 40 percent of the earth's surface. (A message beamed to a satellite from Boston or Montreal could be received in Germany.)

Earth stations transmit the microwave signals through bowl-like antennae. The upward signal is called the **uplink.** The signal relayed back is called the **downlink.** Signals can also be relayed from one satellite to another.

Printing

The third type of people-to-people communication is printing. This includes all systems

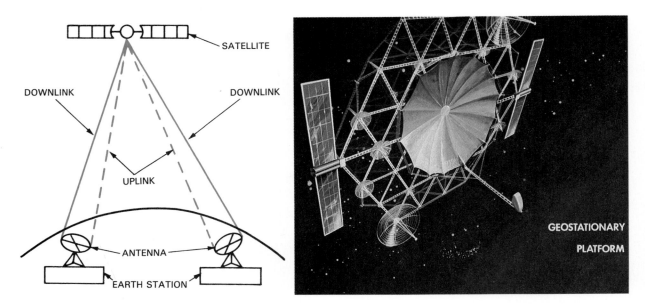

Fig. 10-16. Left. Communication satellite high in space can receive and send signals from earth. Right. A space communication platform of the future. Powerful antenna will be able to link up with small earth stations. You could use it for your own personal communication link!   (NASA)

used to produce letters and pictures on non-treated paper, Fig. 10-17. The printing revolution started with Johannes Gutenberg about 1450. Until that time, scribes produced most books by copying manuscripts one at a time. This process was slow and open to many errors. Gutenberg developed movable type (individual letters on pieces of lead). This allowed the printer to arrange the type for an entire page. Then, a number of copies of the page could be produced. This printing from raised letters on type is called letterpress or relief printing, Fig. 10-18. It is being rapidly replaced by other more efficient processes. One similar process is called **flexography.** This process uses raised type similar to letterpress printing. However, flexography uses a synthetic rubber-like sheet much like a rubber stamp. This sheet is used to print the desired message.

Another more efficient printing process is called offset lithography, Fig. 10-19. It uses a negative (reverse image) of the page to be printed. The negative is placed on an offset plate. The plate has a photographic coating which can be exposed by bright light. The negative allows light to expose the plate where the letters are to be. When the plate is developed, the type to be printed is on the plate.

The plate is placed on a special press. It coats the plate with a liquid (fountain solution). This liquid will not stick where the letters are on the plate. Then ink is applied. The ink will not stick

Fig. 10-17. High speed printing presses produce thousands of newspapers each day. (The Chicago Tribune Co.)

where the fountain solution is. Therefore, ink is only on the type portion of the plate. The plate then presses against a blanket which picks up the ink. the blanket transfers the ink image to the paper, Fig. 10-20.

Other printing processes include intaglio, screen printing, and electrostatic reproduction. Intaglio prints from a recessed image. It is often called an etching and is used for high quality printing. Stamps and paper money are printed by this process.

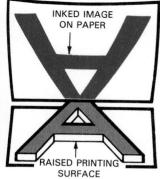
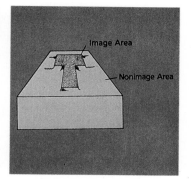

Fig. 10-18. Left. Letterpress printing. Center. Image it produces on paper. Right. Image is made by a raised surface. (Graphic Arts Technical Foundation)

Fig. 10-19. This worker is preparing copy for an offset printing process.
(The Chicago Tribune Co.)

This use of photography is very much different from personal snapshots. These photographs are used to capture a moment of time. They generally are not planned to communicate a message. Therefore, most personal photography is not done for communication purposes. It is done to record a bit of history or an important occasion.

People-To-Machine Communication

People communicate to machines daily, Fig. 10-22. We set controls which "tell" the machine how to operate. We set the thermostat to com-

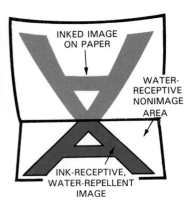

Fig. 10-20. The offset printing process at left produces an image on paper, center from a flat printing surface (right). (Graphic Arts Technical Foundation)

Screen printing, Fig. 10-21, forces ink through a fabric screen to produce an image. Fabrics, posters, and T-shirts are screen printed.

Electrostatic reproduction is rapidly becoming a major printing process. The office "xerox" or photocopier is an example of an electrostatic machine. High speed machines of this type can print several thousand copies an hour fairly inexpensively.

Photographic communication

Directly related to printing is photographic communication. This system uses photographs to convey a message. People on an airliner review emergency procedures with a series of pictures. The message was carefully designed and produced. The pictures tell a story.

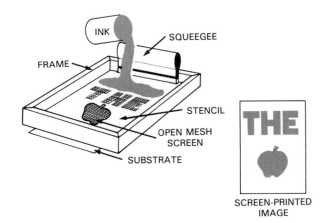

Fig. 10-21. The screen printing process produces an image by forcing ink through openings in a screen carrier.
(Graphic Arts Technical Foundation)

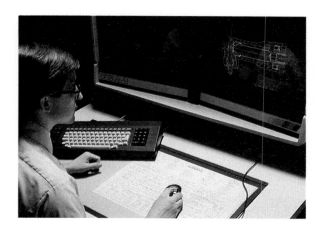

Fig. 10-22. This drafter is using a "mouse" on a menu pad to tell the computer to draw an object.

municate the room temperature we want to a heating/cooling system. We set a speed control system in a car to communicate the speed we want to travel.

Also, we write computer programs to tell the computer what to do. We can tell it to print letters, calculate costs, or perform hundreds of other acts.

Machine-To-People Communication

Directly related to people-to-machine communication is machine-to-people communication. The machine we communicated to through switches and dials responds. It will present us with meter readings, flashing lights, or alarms.

Pilots have many machine-to-people communication systems on the aircraft flight deck, Fig. 10-23. Lights tell them if the engines are running properly. Alarms sound as the plane approaches stall speed. Video screens display the information gathered by the radar system.

In our automobiles, gauges and lights are also used to communicate. The fuel gauge tells the driver when to buy more gasoline. The oil light warns the operator of low oil pressure. A blue light tells the driver that the headlights are on high-beam.

Machine-To-Machine Communication

The most recent communication systems have machines providing information to machines. Computer-aided design (CAD) systems, as shown in Fig. 10-24, help people design parts. They do drafting. The system can then direct machines to produce the part.

Fig. 10-23. This aircraft flight deck is full of dials, gauges, and screens which let the aircraft communicate to the flight officers. (American Airlines)

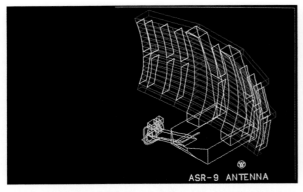

Fig. 10-24. The CAD (Computer Aided Drafting) System shown at top was used to design the radar antenna shown below. (Westinghouse Electric Co.)

Computer-aided manufacturing (CAM) uses computers to directly control machine operations, Fig. 10-25. The computer can direct the machine to run at specific speeds. It can set material feed rates. It can cause the machine to change cutting tools. All this occurs without human action. Even more complex systems totally merge the design and manufacturing activities. These systems are called computer-integrated manufacturing (CIM). These are but a few examples of machines communicating to machine.

THE COMMUNICATION PROCESS

You have now read about many communication systems. Each of these follows a basic com-

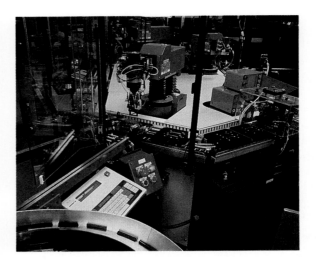

Fig. 10-25. This robotic production line is an example of computer-aided manufacture—CAM. (AMP, Inc.)

munication process. This process, as shown in Fig. 10-26, has four major steps. These steps, which move the information from the sender to the receiver, are:
1. Encoding.
2. Transmitting.
3. Receiving.
4. Decoding.
   At any point along the process the information can be stored. Later it can be retrieved from storage.

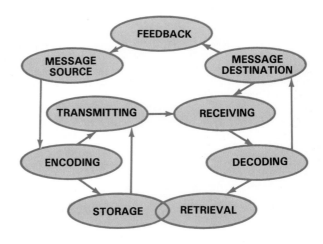

Fig. 10-26. The communication process.

Encoding

Communication involves exchanging information between a sender and a receiver. The sender must first decide what the receiver wants to know, Fig. 10-27. Then the message must be designed. It must be designed to attract the attention of the receiver. Finally, the message must be communicated in a form the receiver will understand. All of these tasks make up **encoding.**

Messages may be encoded in a number of ways. Symbols may be used, Fig. 10-28. A red octagonal (eight-sided) street sign means stop! Two people shaking hands means agreement. A skull and cross bones means poison. All of these are symbols which carry meaning.

Other messages may be written. The information may be communicated using language. This book is communicating a message in this manner.

Fig. 10-28. This symbol is used to identify restroom and parking areas which are designed for handicapped people.

We have already presented the basic ways of transmitting (sending) the message. These are by graphic means or wave transmission. The message may be printed on paper or carried by a series of photographs. These are the graphic communication technologies. Electronic (or wave) messages may be carried on sound, light, or radio waves, Fig. 10-29. We may use radio or television transmitters to send our information. Or, light produced by a laser may carry our telephone conversations. A public address system may use sound waves to broadcast our ideas.

Fig. 10-27. This designer is developing the message and layout for an advertisement. (Ohio Art Co.)

Transmitting

Once the message is encoded it must be delivered to the receiver. Remember, we are talking about communication technology. Therefore a technical means (machine, equipment) must be used.

Receiving

The transmitted message must be received, Fig. 10-30. The message must arrive at a desired location. It must be available to the receiver. The message sent by the transmitter is often in a special form. It may be a series of electrical pulses on a wire. Or, it may be radio waves which vary in strength (AM radio) or in frequency (FM radio). Or, it may be in digital code.

Often receivers are electronic devices. They change the transmitted code back into a form people can understand. They may change radio waves into sound waves. Or, they change digital code into printed words.

Fig. 10-29. This scene shows the first stage in transmitting a message using electronic media. (American Petroleum Institute)

Fig. 10-30. This drafter is receiving information using a common communication system, the telephone system.
(American Petroleum Institute)

Decoding

The final act is putting meaning to the message. The people receiving the message must take action. They must read the printed word, listen to the broadcast, or watch the television program. But that isn't enough. They must place their meaning on the message.

For example, you might hear someone shout "fore." First, you must decide if the word is "fore" or "four." Then, you must put it in context (compare it with the situation). If you are on a golf course you should become alert. "Fore" means someone is hitting a golf ball. If you are cooking at a fast food outlet you may take other action. The manager may have told you to make "four" hamburgers.

A message is only effective if it is received. Time can be wasted designing, producing, and sending messages that are never received. How many tornado, flood, and hurricane warnings are transmitted but ignored. Many no smoking signs go unnoticed. Highway speed limit signs are viewed with indifference. Health warnings on cigarette packages are disregarded.

Storage/Retrieval

At any point in the communication process, the message may be stored. It can be recorded on video and audio tape, Fig. 10-31. Or, it can be stored on computer discs. Or, the printed message can be placed in a warehouse.

Fig. 10-31. This customer is selecting a video disc.   (Rohm and Haas)

Later the message is taken out of storage. It re-enters the communication process. Most television programs are taped (stored) for later broadcast (retrieved). Books are stored in libraries. The reader must retrieve and decode the messages.

PRODUCING COMMUNICATION MESSAGES

Formal communication messages are carefully planned and produced. Their production involves four major steps. These, as shown in Fig. 10-32, are:
1. Designing the message.
2. Preparing to produce the message.
3. Producing the message.
4. Delivering the message.

These steps apply to all major communication activities. Let's look at how these activities apply to three specific industries: publishing, filmmaking, and broadcasting. You will see that there are some things that each group does alike. Other activities belong only to a specific industry.

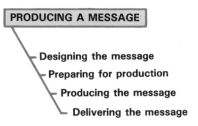

Fig. 10-32. Steps in designing a communication message.

Producing Published Messages_____

Publishing is all activities which produce a printed message. This includes newspapers, magazines, books, greeting cards, flyers, etc. Publishing is part of a larger industry called printing. In addition to publishing, printing includes the production of forms, stationery, and other "nonmessage" materials.

Designing published messages

Mass communication messages are designed to reach large audiences. They are planned to give them information or cause them to take action. But there are many audiences. Therefore, the designer must gather information about a specific audience he or she wants to reach. In particular, the following questions must be answered:
1. Who is the audience (young people, sports fans, business leaders, senior citizens, etc.)?
2. What gets the audience's attention (words, pictures, comic presentations, etc.)?
3. What does the audience value (status, financial security, fun, freedom, power, etc.)?

Gathering this information is called **audience assessment.** The information provides the base for designing the message. It gives the designer guidance and direction to complete several specific tasks.

The first design task involves selecting a **format.** This is planning for the physical size and shape of the message carrier. Will it be an 8 1/2 x 11 in. flyer? Or could it be a large billboard on a busy highway? Will the newspaper be a tabloid, (smaller size) or regular format?

Next, the designer gathers information. The message may be designed to sell products (an advertisement), Fig. 10-33. Then information

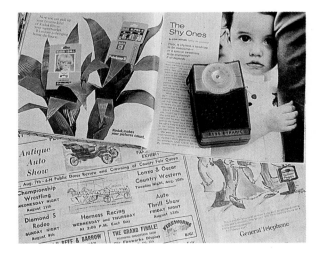

Fig. 10-33. These are examples of advertisements designed to sell products.

about the product's features and operation are needed. This can be matched with audience information to determine what to say. The designer then *writes copy* for the message. It is the story that the media is to carry.

Newpaper reporters also gather information. They interview people associated with a story. Then they select and combine information to produce interesting copy.

Finally, the designer selects illustrations to support the copy. These drawings and photographs will attract attention and add interest to the message. They may help the news story convey information.

Preparing to produce published messages

The format, copy, and illustrations must be brought together. They must be placed in a pleasing arrangement.

In designing advertising, designers first sketch out their ideas. First, **roughs** are prepared to show some basic ideas, Fig. 10-34.

Fig. 10-34. This artist is preparing "roughs" for an advertisement.
(American Petroleum Institute)

The better ideas are converted into **refined** sketches. These show more detail and thought. Finally, **comprehensives** are prepared. These are more complete sketches for the message. They describe the arrangement of the elements—type, illustration, and white space. Type size and style are called out (written on the sketch).

Likewise, the newspaper has a layout which is designed for the audience. "USA Today" uses a colorful format with many photographs, charts, and graphs. The "Wall Street Journal" caters to (pleases) its audience with a large amount of text.

Producing published messages

The message is now ready to be produced. The advertisement, magazine, or newspaper is ready to be printed. The first major task is to schedule the production through the plant. Resources must be allocated (given) to each production task.

Then, the comprehensive is used to set type, size illustrations, and prepare photographs. These components are then attached to a large sheet of paper. This is called a **layout** or a pasteup.

The pasteup is used to make printing plates or silkscreens for the production processes. These items are called **image carriers.** They carry the image from the press to the paper.

Once the large carriers are produced, the communication product must be printed. Ink must be applied to the substrate (paper, plastic, foil, etc.). As described earlier, this can be done by one of five basic processes:

1. Letterpress (or relief) printing. Printing from a raised surface.
2. Offset lithography. Printing from a flat surface.
3. Intaglio (or gravure) printing. Printing from a recessed surface.
4. Screen printing. Printing by forcing ink through openings in a screen.
5. Xerography. Printing using electrostatic (charging a drum electrically) means.

Delivering published messages

Delivering a printed message uses some standard distribution methods. Subscribers may receive the message in newspapers and magazines. Books are available in stores and libraries. Billboards, posters in store windows, and mailed flyers are still other distribution techniques.

The task is to select a distribution method that will reach the identified audience. Using "Sports Illustrated" magazine to reach large numbers of older women would be unwise. But, a message in "Seventeen" magazine will reach many school-aged young women.

Producing Film Messages

Film messages are photographs and transparencies (movies, slides, and filmstrips) designed to present information or entertainment.

Designing film messages

Like published works, they are based on the results of audience assessments. Moviemakers have a good idea what teenagers will go to see. It is greatly different from what grandparents feel are "good movies."

The first design decision concerns format. Will the message be presented by a motion picture or a filmstrip? How long should the presentation be? Should it be color or black and white? Answering these questions will establish the format.

Next, research must be done. Information about the subject must be developed. Educational films require much research. A good example of this type of film is the "National Geographic Presents" series.

Entertainment films often take less research. Sometimes a film is based on a novel. Then the author has already completed the research.

The third step is *scriptwriting*. Most films use one or more actors to portray a story or present information. Their actions and words must be written down. This is called the script, Fig. 10-35. The script carefully describes all events in the film.

Stage sets are designed and built to support these actions, Fig. 10-36. Also, many times on-site filming locations are selected and used. The sets and locations provide the realism for the message.

Various Voices off: *H'ya, Buck! . . .Howdy, Buck! . . .How's things in Bisbee, Buck? Have a good trip?*

Meanwhile the SHOTGUN GUARD, who has guarded the treasure box from Bisbee, jumps down to the sidewalk.

SHOTGUN GUARD: *So long, Buck!*

Men begin unhitching the horses. BUCK acknowledges the cheery greetings as the WELLS FARGO AGENT in Tonto pushes his way through the crowd.

WELLS FARGO AGENT: *Howdy, Buck. Got that payroll for the mining company?*

Buck kicks the box which is under his seat.

BUCK: *She's right here in this box.*

The WELLS FARGO AGENT climbs up to the top of the coach, calling to a colleague as he does so.

WELLS FARGO AGENT: *Give us a hand with this box, Jim.*

BUCK: *Jim, I'll pay you that $2.50 when I get through.*

JIM: *Okay.*

The two agents get the box down and carry it off between them — BUCK looks over his shoulder to the other side of the coach.

BUCK: *Now you kids, get away from them wheels!*

He starts to get down and calls out to the men who are leading the horses away.

BUCK: *Well . . . sir, we ran into a little snow up there, quite bad, so you fellers better prepare for a good frost.* He jumps down and disappears round the side of the coach. The Tonto Hotel is seen on the other side of the road.

Medium shot of the stagecoach as BUCK comes round to open the coach door.

Fig. 10-35. This is a page from the script for STAGECOACH, a John Ford and Dudley Nichols film. The script was printed by Lorimer Publishing, London, England.

Fig. 10-36. This outdoor set can be used in making movies and television shows. Notice the walkways at the top of the "buildings."

Preparing to produce film messages

The production of film messages must be scheduled. The efficient use of resources must be planned. Then, the cast and production crew must be hired.

The director must stage the various scenes. He or she decides how each scene will be shot. Placement of cameras and lights are considered. Also, the director determines how to use extras (people who add interest to the scene). The camera, lighting, wardrobe, and set crews work closely with the director.

The cast learns lines and movements from the script. Then they come together to *rehearse* (practice). Changes are made in the script and the staging until the director, cast and crew are happy with the production.

Producing film messages

After the final rehearsal, filming can start. The actors complete each scene they rehearsed. (The scenes are seldom shot in the order outlined in the script.) Wise use of resources (stage time, natural light, location availability, etc.) dictate the proper order to shoot the scenes.

Crew and cast may shoot each scene a number of times to develop several different effects. Later, the film is cut and spliced to combine the various scenes. They are edited into a final product.

Distributing film messages

Most movies are first distributed to theaters. Special companies schedule the films and col-

lect royalties for the film producers. Other films are made for television. They are distributed to the television stations.

In the past few years films have had a second channel of distribution. After they are shown in theaters, they are offered on video tapes. Individuals may buy or rent the movie to view at home.

Producing Broadcast Messages

Broadcast messages are communications carried by radio and television stations. These include entertainment, news, sports events, and advertisements.

Regular programming follows the basic steps outlined for films. Scripts are prepared, crews and actors are hired, the production is rehearsed and filmed, and the product is edited into its final form.

The major difference is in the way they reach the viewer. Film producers expect people to pay to see the program in a theater. Broadcast programming sells advertising time to enable "free" delivery of the product. The word, free, is in quotes because it is NOT free. We pay for radio and television programming everytime we purchase an advertised product. Likewise, we pay for part of the production cost of magazines. and newspapers with our subscriptions. But we also pay the rest of the costs through the advertisers.

Broadcast advertisements follow the design steps used for print advertising described earlier. The only difference is in the layout. Radio advertisement uses a script to "lay out" the ad. Television uses a **storyboard,** as shown in Fig. 10-37. The storyboard shows each shot for the advertisement. Also, the script is included.

Producing broadcast advertising follows the film model. The director and actors rehearse the advertisement. Filming and editing then take place. Finally, time is purchased to present the advertisement of the broadcast station.

Most 60 second advertisements will be shot a number of times (sometimes into the hundreds) to get just the right effect. The ad is a sponsor's way to cause people to buy his or her

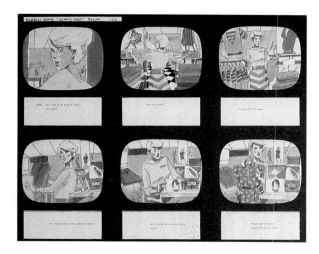

Fig. 10-37. This is an example of a storyboard for a television commercial. (Clorox Co.)

product. A great deal of money is spent to produce and air the short message. Therefore, they want it to be nearly perfect.

SUMMING UP

Communication technology is part of each person's life, Fig. 10-38. We use graphic and electronic means daily to send or receive information. These messages are carefully designed to inform us or to cause us to take action. Without communication technology we would know little about the world around us. We would live a very sheltered life.

KEY WORDS

These words are used in this chapter. Do you know their meaning?

Audience assessment, Audio and video recordings, Broadcast messages, Carrier, Communication, Communication satellite, Communication technology, Comprehensives, Decoding, Digitized, Downlink, Electromagnetic radiation, Encoding, Film messages, Flexography, Format, Graphic communication, Layout, Media, Photographic systems, Printing systems, Published messages, Receiver, Roughs, Storyboard, Telecommunication system, Transmitter, Transmitting, Uplink, Wave communication.

TEST YOUR KNOWLEDGE
Chapter 10

Do not write in this text. Place answers to test questions on a separate sheet.
1. Indicate which of the following are communication:
   a. Thinking about what you will do today.
   b. Talking to a friend.
   c. Listening to music.
   d. Tapping a friend on the shoulder.
   e. Tasting your dessert.
   f. Whistling.
   g. Drawing a design for a poster.
2. What makes communication a technology?

Fig. 10-38. Can you see any communication technologies in this photo? How about the records, television, and newspaper? (Rohm and Haas)

Match the terms to the definitions:

3. ____ Uses printed words and pictures.
4. ____ Uses electromagnetic waves to carry messages through space.
5. ____ Waves that travel at 750 mph at sea level.
6. ____ Uses light to send and receive messages.
7. ____ Uses echoes to collect information.

A. Sound waves.
B. Fiber optic communications.
C. Sonar.
D. Graphic communications.
E. Wave communication.

8. List and describe the four types of communication.

Indicate the type of communication system used in each situation described:

9. ____ Listening to a tape player.
10. ____ A computer which can operate a robot.
11. ____ Clock chimes sounding the hour.
12. ____ Setting the alarm on a clock radio.
13. ____ Thermostat which controls operation of a furnace.

A. People-to-people communication.
B. People-to-machine communication.
C. Machine-to-people communication.
D. Machine-to-machine communication.

14. List the steps that move information from a sender to a receiver.
15. _____ means to give a message received a meaning.
16. List the four major steps in producing communication messages.
17. Name the five printing processes.
18. _____ _____ are photographs and transparencies designed to present information or entertainment.

19. A script is to a radio advertisement what a _____ is to a television commercial (advertisement).

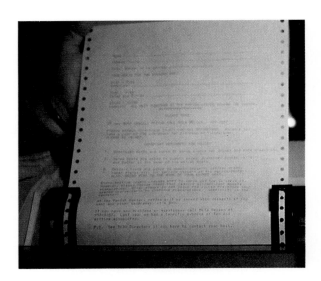

Today, it is important to learn how to "talk" with computers. This company executive is getting information from an oil company operator. (American Association of Blacks in Energy)

A computer talks to other machines such as a printer in the form of electrical impulses. The message is printed up much like the communication from a typewriter.

APPLYING YOUR KNOWLEDGE

Introduction

We communicate daily. Some of our communication is verbal (speech). People talk to people. Some is visual. We write messages for others to read.

Often we use a technological device to help us deliver our message. We may use printing presses or cameras to produce two-dimensional media, the printed page, or a photograph. Other times we use electrical signals or electromagnetic waves. We use the radio, television, telephone, teletype, or telegraph. These devices move a message from the source to the receiver.

This activity will let you build a telegraph system. Then you will use this technological system to convey a message.

Equipment and Supplies

2 — 3/4" x 3 1/2" x 5" pine boards
2 — 5/8" x 5 1/2" x 28 gauge sheet steel
6 — 1/2" brads
1 6d box nail
12 — 3/8" x No. 6 pan head sheet metal screws
2 — Mini buzzers — Radio Shack No. 273-055 or equivalent
4 — Quick wire disconnect — Radio Shack No. 274-315 or equivalent or, 6 — Fahnestock clips
1 — 9 V battery holder
25 ft. 3-conductor wire or 75 ft. of single conductor wire.

Safety

- Be careful when using tin snips. Keep your free hand well away from the cutting edges.
- Edges of sheet metal may be sharp or jagged. Handle sheet metal with caution.
- In this activity you will use only a 9-volt power supply. There is no danger from such low voltage. Still, it is possible to get an uncomfortable shock under some conditions. Do not touch bare wires or terminals.
- Always work carefully with tools and materials. Don't clown or attempt practical jokes. This could cause injury to yourself or others.
- If you are uncertain what to do, ask your instructor. Don't guess!
- DO NOT attempt to connect the Telegraph Stations to any other power source than the 9-volt battery.

Procedure

Your teacher will divide you into teams of four students. Each team will be split up into two groups:

Group A will build Telegraph Station No. 1.
Group B will be responsible for building Telegraph Station No. 2.

How to build

Each member of a group should:
1. Carefully study the plans for the telegraph set. (Refer to Fig. 10A for Group A and Fig. 10B for Group B.)
2. Select a piece of lumber for the base of the set.
3. Lay out the location of the key, buzzer, and wire disconects or Fahnestock clips.
4. Select a piece of 28 gauge sheet steel to make the telegraph key.
5. Lay out the pattern on the metal, Fig. 10C.
6. Cut the metal to size.
7. Punch the two screw holes.
8. Bend the metal on the two bend lines to form the key as shown in Fig. 10D.
9. Drive the 6d box nail in the correct position. Allow 1 in. of it to remain above the surface of the wood.
10. Attach the telegraph key, buzzer, and wire disconnects.
11. Wire the telegraph station according to the original drawing.

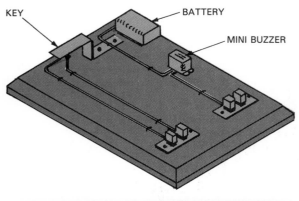

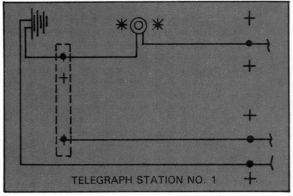

Fig. 10A. Group A will construct this telegraph station. Carefully study all of the drawings first.

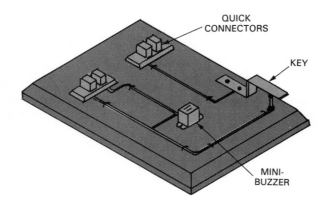

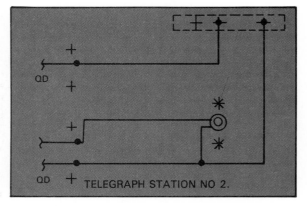

Fig. 10B. Group B will construct this station. Note that it is different from the one being built by Group A.

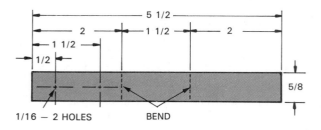

TELEGRAPH KEY PATTERN

Fig. 10C. Pattern for the telegraph key.

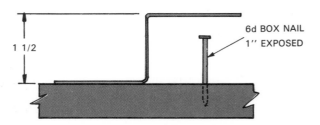

Fig. 10D. Telegraph key should be bent to this shape.

Sending the message:

1. Wire Telegraph Station No. 1 to Telegraph Station No. 2 as shown in Fig. 10E.
2. Complete one of the two following assignments using the telegraph system to communicate messages between Group A and Group B.

Assignment 1:
Group A and Group B will each take a short sentence from this chapter of your text. Convert (encode) the sentence to Morse code. The code is shown in Fig. 10-11. Then Group A will transmit their sentence to Group B. Group B will receive

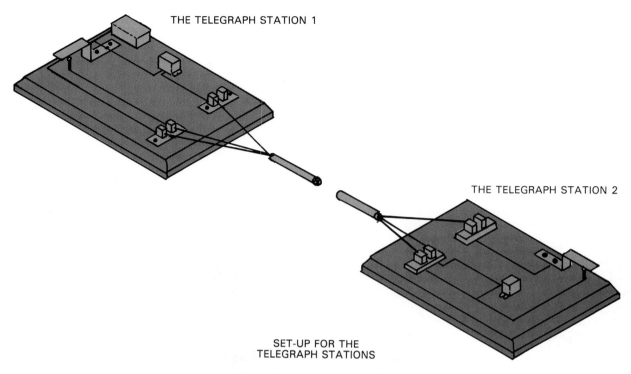

THE TELEGRAPH STATION 1

THE TELEGRAPH STATION 2

SET-UP FOR THE
TELEGRAPH STATIONS

Fig. 10E. Wire the two telegraph stations together as shown.

and decode the message. Then Group B will send their message to Group A where it will be recorded and decoded. Check each "received" message for accuracy.

Assignment 2 — Moving Through a Maze.

1. The teams will develop a code that will tell a person how to move through the maze. Typical comands: "Go left. Go right. Go back. Stop. Pass the first opening. Move to the second opening, etc.

2. Group A should lay a piece of tracing paper over Maze Drawing No. 1 (Fig. 10F) in their book and solve the puzzle. NOTE: Your teacher may provide you with enlarged copies of the maze so you can write directly on them.

3. Group A will then develop a series of commands to help Group B solve the puzzle.

4. Group B will lay a piece of tracing paper over Maze Drawing No. 1 in their book.

5. Group A should use the telegraph system to transmit the commands for Group B to move through the maze.

6. Group B should receive and decode the commands. Then they should draw a path through the maze on the tracing paper. This path should be based on the directions sent on the telegraph.

7. The team should compare the accuracy of the message sent and the message received.

8. Group B should complete steps 3 through 7 to send their maze directions (Fig. 10G) to Group A.

Assignment 2 Researching and Writing a Report

In this assignment you will learn how electric current is able to move through the telegraph system. You will find out how current makes the mini-buzzers work. You will then produce a report on what you have learned. You will need pencil or pen, a notebook, a computer, and a word processing program.

1. The team will use books provided by your instructor or your media center to learn how

electrical current travels through conductors. Also, study how this current can be controlled.
2. Record this information in the notebook.
3. Write a short report on your findings.
4. Use the computer and the word processing program. Prepare the report and file (save) it on a disk.
5. Print a copy of the report. Give it to your instructor.

START

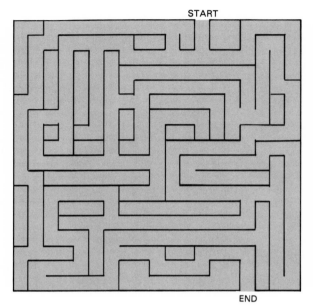

END

Fig. 10F. Group A will solve and produce a code for running this maze.

START

END

Fig. 10G. Group B maze. The group will first solve the maze and then produce a code for running it.

Chapter 11
Transportation Systems

The information given in this chapter will help you to:
- ☐ Define transportation.
- ☐ List the modes (ways) of travel used today.
- ☐ Trace the development of transportation.
- ☐ Tell what inputs are needed for a transportation system.
- ☐ Describe the various parts that make up a transportation system.

Transportation is the movement of people and goods from one place to another. You may wonder how it got the name. It comes from two Latin words, "trans" and "porto." "Trans" means "across." "Porto" means to "carry."

Transporting requires vehicles. There are many types. The ones we think of first are automobiles, buses, trucks, and airplanes. Other types of transport never leave a certain location. They are called on-site transport. Can you think of such a transport vehicle? What about an escalator?

Transportation is important to our way of life. Without it travel distance would be limited to how far we could walk. All our food would have to be grown nearby. Our clothing would be made from materials grown or found where we live, Fig. 11-1.

What about housing? Could we find enough wood, stone, and other building material with-

out transport? Indeed, transportation is a necessary part of our lives, Fig. 11-2.

MODES OF TRANSPORTATION

There are different ways of traveling or moving material. We call the ways **modes.** For ex-

Fig. 11-1. Without transportation we might have to go back to a time when people made their own fabric. This woman demonstrates how pioneers spun yarn from wool and cotton.

A B C

Fig. 11-2. Modern conveniences are possible through transportation. A—Grocery stores have food grown and packed in a foreign country. B—Modern shopping mall has clothing and many other products. C—Building site. You see steel, plywood, and concrete. Were any of these materials made locally or on site?

ample, you may ride your bicycle to school. But you might ship a package on a plane. You might travel in a boat to cross a lake or river. Modes of travel include:

1. Land—using cars, trucks, trains, pipelines, escalators, and conveyors.
2. Water—using boats, ships, and barges.
3. Air—using airplanes, helicopters, and, sometimes balloons.
4. Space—using rockets, other space vehicles.

Degrees of Freedom

Did you ever hear anyone say: "you can't get there from here"? As funny as the saying seems, you can't travel or ship goods everywhere you might want by every mode. For example, you have to go where a river takes you. You have to follow a certain road. Every mode of travel has some ways you cannot move. We call them restrictions. Some modes offer more freedom than others, Fig. 11-3.

Consider these restrictions: Escalators can only take you up or down. (The word for this is uni-directional, meaning "one direction.") But an elevator can go up or down. It is two-directional. So are trains and pipelines. Cars, buses, planes and other vehicles can be steered.

They can move forward or backward, left or right. Planes can go every direction. Plus, they can go up and down. Such vehicles are "multi-directional."

Pathways

Pathways and safety needs restrict where a vehicle can travel. *Pathways* are the routes different transport vehicles must use. They have three purposes:

1. To support the vehicle. Roadways and rails support land transportation vehicles, Fig. 11-3, views B and C. They provide a smooth, durable surface over which the vehicle can travel. Waterways support ships. The water must be deep enough to float the water vehicle and its load. Look at Fig. 11-3A again.
2. To allow vehicles to reach their destination. Natural barriers must be removed. Rivers must be bridged. Mountains may be tunneled. Surfaces must be strong so vehicles do not become stuck.
3. To allow vehicles to move without hurting people or damaging property. Imagine what would happen if vehicles could move anywhere. No one would be safe!

A   B   C

Fig. 11-3. A—This ferry must travel only in a river channel or back and forth in a small body of water. B—Rails support trains and restrict where they can move. C—Do you see how important roads and streets are to a large city?   (American Petroleum Institute and CSX Corp.)

Some pathways restrict vehicles more than others. Rails, moving sidewalks, elevators, and piplines are very restrictive. Trains must stay on the tracks. Material in pipelines is closely controlled. Usually, it is underground.

Cars, trucks, and buses can move more independently. True, they must follow the road. But there is often a choice of routes.

Do you think that waterways are more restrictive than roads? You are right. There are fewer rivers and canals than roads. There aren't as many places to go. Ocean shipping lanes are not as restrictive. Ships can often move out of shipping lanes without any serious consequences. But on inland waters there is always a danger of running aground. Water vehicles must stay in the channel (deep part). Locks must be installed if water levels of the waterway change. See Fig. 11-4.

Aircraft must travel in marked airways. Craft moving in one direction must fly at a different altitude (height) to avoid midair collisions.

Land Transport

All types of transportation that move on or beneath the earth's surface are known as land

Fig. 11-4. Locks help water vehicles move between waterways that are at different levels. This is the Bay Spring Lock in the Tennessee—Tombigbee Waterway.   (National Park Service, Natchez Trace Parkway)

transportation, Fig. 11-5. They all move over constructed pathways. All systems are one or two types:

1. Point-to-point. Railroads, pipelines, and conveyors, are this type. So are on-site systems. These include elevators, moving sidewalks, and escalators.
2. Steerable. Cars, buses, trucks, forklifts, and motorcycles are examples.

Railroads

Railroads were important long before any steerable vehicles were built. They were vital to the settling of the West. They did more than carry pioneers to new homes. More important, they opened up eastern markets for the livestock and grain produced on the frontier. They brought back manufactured goods that the pioneers badly needed.

Fig. 11-5. Land transport vehicles move over constructed paths. Trains travel over rails supported by wood crosspieces called ties. This train is hauling a cargo of 100,000 tons of oil shale at Rifle, Colorado. Cars and trucks travel on highways.   (Paraho Development Corp.)

Trains are an efficient way to move people and goods. The tracks and the vehicles are built to carry heavy loads. They can carry many passengers and huge amounts of cargo. And because their pathways are not crowded, they never get caught in traffic jams!

Hauling material

Material-handling trains are either unit trains or freight trains. The difference is in their cargo and how the cargo is delivered.

Unit trains carry only one kind of material. They move the cargo to the same destination time-after-time. For example, coal could be moved from the mine to a distant power station.

Freight trains are made up of individual cars carrying many different products and materials. A freight train traveling from Chicago to the west coast will drop off cars and pick up others along the way.

Moving people

Two classes of travelers use passenger trains. Commuters take them to and from their jobs. Long distance travelers may ride to faraway cities or even coast to coast.

Rolling stock

A train's cars are called **rolling stock.** The name applies to engines and maintenance vehicles too.

In the past, steam powered the engine or locomotive. Today's engines are diesel-electrics. First, the diesel engines generate electricity. Powerful electric motors use the electricity to drive the wheels.

Railroad cars are designed for hauling different kinds of freight. Boxcars haul loose or packaged freight. Other cars are designed for special needs. Tank cars haul liquids, for example. Gondola cars haul coal. Maintenance vehicles repair right-of-way, cut brush, and lift derailed cars.

Steerable transport

Cargo and people can be moved by way of expressways, highways, streets, and even inside buildings, Fig. 11-6. Schedules are flexible. The driver has great independence. Travel is possible day or night.

| A | B | C |

Fig. 11-6. Cargo and people are moved by steerable vehicles on various pathways. A—On streets. B—For harvesting. C—Inside buildings. (Caterpillar, Inc.)

Steerable transport includes:
1. Personal. People of the United States and Canada often travel by automobile. They prefer the freedom and convenience.
2. Commercial. Commercial transport means hauling for hire. Buses and taxis carry people for hire. The passengers pay a fare to ride. Transport may be short haul, that is, a few miles. Long haul means between distant cities or across the continent. Large freight vehicles are known as semi-trailers or "18 wheelers." They are a combination of a trailer and a tractor, Fig. 11-6A.

A combination land or water vehicle is known as a "ground effects" machine or *hover craft*. A fan in the vehicle creates a thin cushion of high pressure air. The machine moves about on this cushion of air, Fig. 11-7. The first commercial use was in 1962. The British began using it for water transport for short distances. The U.S. Navy is now experimenting with it.

Pipelines and conveyors

Some materials can be moved through **pipelines** or on conveyors. Pumps suck or push the material in a pipeline. Petroleum and petroleum products, natural gas, and coal are often moved this way. So are grain, gravel, and wood chips. Fig. 11-8 shows a pipeline under construction.

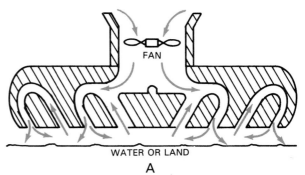

A

B

Fig. 11-7. A—How hovercraft works. Vehicle floats on high-pressure air bubble. Fan at top compresses air and sends it to bottom of craft. B—A hovercraft developed by the U.S. Navy. It was tested in the Arctic. One day it may also be used for oil exploration there. (Standard Oil Co.—Ohio)

Fig. 11-8. Pipeline under construction. Pipelines are efficient movers of materials whether solids, liquids, or gases. (USX Corp.)

Pipelines have advantages over other methods of moving material.

1. The material moves but the vehicle (pipe) stands still. It lasts longer.
2. Most pipelines are buried. They conserve valuable land. There is no congestion or danger of accidents.
3. There is little danger from thieves. Neither is there much chance of damage to or contamination of materials.

Conveyors are stationary, built-in structures that move materials and products. They are often used in manufacturing, Fig. 11-9. They move materials:

1. Along a production line during assembly.
2. From mine shafts or pits to processing operations.

Conveyors are also "people movers." They can be made to transport people from one part of large buildings to another (moving sidewalk). Conveyors may be driven by electric motors. They can be made up of continuous belts or chains.

Water Transport

Nature provides the paths we call waterways. They usually do not have to be constructed. But, sometimes the waterways must be made deeper. Human made canals may be dug between natural bodies of water, Fig. 11-10.

Water transport is generally cheaper than land transportation. However, it can be used only where rivers, lakes, and other bodies of water are *navigable*. That is, they must be wide enough and deep enough for heavily loaded watercraft to travel on them, Fig. 11-11.

Water transport vehicles

A number of vehicles are used for *inland shipping* (not on the open seas). **Barges** are large floating cargo boxes. They have no engines. **Tugboats** or towboats are small vessels. They are the "locomotives" of the water. They push or pull the barges. Sometimes they also move large vessels in close quarters. **Ferries** move people and vehicles across bodies of water.

A   B   C

Fig. 11-9. Conveyors move products short distances. A—New auto bodies move down an assembly line on a conveyor. (Chrysler Corp.) B—A crane moves on an overhead track. It moves heavy materials from one spot to another. It can move anywhere in the building—left, right, up, or down.   (USX Corp.) C—A conveyor belt moves crushed rock at a quarry.

Fig. 11-10.   This canal connects two navigable rivers in Mississippi. It is an important route for barge traffic. It took many people and millions of dollars to dig it. Locks had to be installed to lift boats and barges to different water levels along the way.

Fig. 11-11. Barge transport is a familiar sight on canals and rivers. Notice the deck which is meant to protect the cargo, possibly grain. (U.S. Park Service, Natchez Trace Parkway)

Freight and people are carried by larger ships across oceans to and from other countries. Those carrying people are called **ocean liners.** **Freighters** carry products and solid materials. **Tankers,** Fig. 11-12, carry liquids like petroleum and chemicals.

Fig. 11-12. A tanker pushes through ice in Chesapeake Bay. It was delivering oil products to Baltimore during the winter as this photograph was taken.   (Exxon)

Ocean pathways

There are no "highways" on the open seas. However, ships do keep to regular routes over the water. These routes are known as **sea lanes** or shipping lanes. Ships travel along these routes. They stay on course with compasses and other navigation tools.

Air Transportation

Aircraft include all vehicles that travel within the earth's atmosphere. Airplanes are the most-used aircraft.

Airplanes are a faster way to travel. They move cargo as well as people.

Commercial aviation includes all air transport done for profit. The aircraft are very large, Fig. 11-13.

Private planes make up a second transportation group. This group is called *general aviation.* The planes, Fig. 11-14, are generally smaller than commercial craft.

The military also has a fleet of aircraft. They have many kinds and sizes of airplanes.

Fig. 11-13. A commercial airliner during takeoff. It carries passengers between distant cities at high speed.   (United Airlines)

Fig. 11-14. Small aircraft are for personal use or for commercial flying. This one flies vacationers to remote lake regions of Canada.

Special aircraft

Aircraft have been developed for vertical takeoff and landing (VTOL). Their main advantage is that they need only small landing space. (Usually, this landing space is called a "pad.") Fig. 11-15 shows two types of aircraft. The helicopter is a common type of vertical takeoff and landing craft. It has two rotors. One provides lift and forward motion. A second, smaller, rotor is located in the tail section. It provides stability so that the direction of flight can be controlled by the pilot. It acts like a rudder. This rotor also prevents the helicopter from spinning around.

How the helicopter flies

The blade angle of the main helicopter rotor can be changed. When lift is needed the angle changes to "bite" the air. At the same time, the engine speeds up so the rotor turns more rapidly.

Another change in the blade pitch enables the rotor to "pull" the helicopter forward through the air. For landing, the rotor blade pitch changes again. At the same time, engine speed drops.

Helicopters are the "workhorses" of the air. They are used as air ambulances, to haul materials, as taxis, to spray crops, to clock traffic speed, to patrol power lines, and as airborne cranes in construction.

The XV-15 is an experimental plane built by an American helicopter manufacturer. It has been undergoing testing. Its rotors tilt upward for takeoff. Then they drop to a horizontal (level) position for regular flight. This makes them much faster than helicopters.

Airships (like balloons, lighter than air) are another kind of air vehicle. They are more limited in use. The military have used them as a way to watch enemy forces.

Skyways

Like the oceans, the skies have no highways. But, to be safe, aircraft must follow certain routes. In addition, air lanes are often at different heights. (For example, planes flying west may travel at a different height than planes flying east). A federal agency controls all air traffic.

Space Transportation

Space travel is very new. It began in 1957 when Russia launched a satellite into an orbit around the earth. Later American astronauts

A

B

Fig. 11-15. Two types of craft that can take off and land vertically. A—Helicopters have two rotors. One is always horizontal. The other is always vertical. B—NASA/Army research aircraft. Its rotors can be tilted upward for takeoff. Then they tilt forward and propel the craft like an airplane. (NASA)

traveled to the moon. Space vehicles are called **rockets.** Because there is no oxygen in outer space, rockets carry their own oxygen. It is used to burn the rocket fuel.

Space transportation includes unmanned flights and manned flights. Unmanned flights have rockets traveling far into outer space. They are exploring the universe, Fig. 11-16. Cameras aboard the space vehicles send back photographs of unexplored space. Manned flights have taken astronauts to the moon. The space shuttle ferries humans between space stations and earth. Communication satellites orbit the earth to transmit messages.

Intermodal Travel

When we use more than one mode of transportation it is called **intermodal transportation.** See Fig. 11-17. In our lifetimes, most of us will travel by land, water, and air. When we fly we must take a bus, train, or automobile to the airport. If we take a cruise, we travel by car or plane to the port terminal. During the cruise we may visit other ports. We will use other modes of travel. We may even ride in vehicles drawn by humans.

SUPPORT SYSTEMS— STATIONS AND TERMINALS

Stations and terminals are structures that shelter transportation activites. Every mode of transportation has them. They are needed to:

1. Allow easy access for intermodal travel. Buses and automobiles bring passengers to and from airports. There must be facilities for unloading. Baggage must be moved. Travelers need information, tickets, food, and restrooms. See Fig. 11-18. Dockside facilities provide needed services for cargo handling. Vehicles and cranes move seagoing containers from ships. They are transferred to and from land transportation. Loose materials such as grain must be transferred. There must be unloading facilities where trucks can deliver and collect the material. Conveyors move the material to and from barges and railroad cars. Escalators, elevators, moving sidewalks, stairs, and conveyors are used at terminals. They help travelers reach airplanes, buses, and other modes of transportation. Goods can be moved by steerable

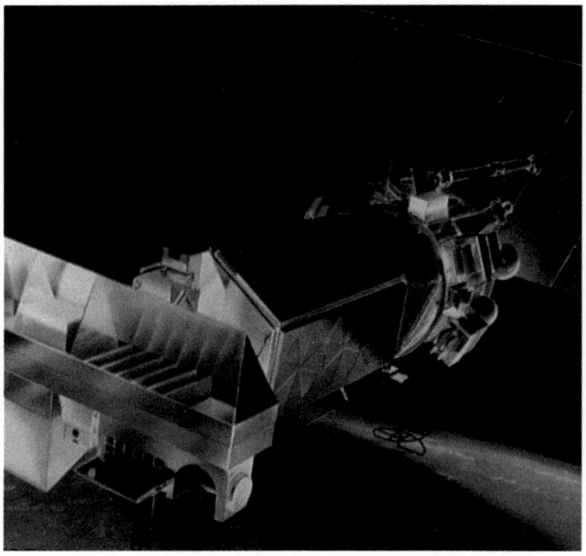

Fig. 11-16. This space satellite takes pictures from 600 miles above earth.   (Comsat)

Fig. 11-17. Intermodal transportation means use of more than one type of transport. Three are shown here: barge and tug (water), train (land), and truck (steerable, land).   (CSX Corp.)

A
B

Fig. 11-18. Terminals provide essential services for travelers. A—Airport personnel sell tickets and give out information. B—An old abandoned depot once served railroad travelers in Mississippi (U.S. Park Service, Natchez Trace Parkway)

vehicles such as forklifts. See Fig. 11-19.
2. Provide security and shelter. Terminals provide warmth and safety for people. Material is protected from weather and theft.
3. Provide vehicle maintenance, Fig. 11-20. Trucks, airplanes, ships, and other vehicles need maintenance and other types of service. This keeps them in good operating condition. Lubricants must be changed, vehicles refueled, washed, and tires replaced.
4. Store transportation vehicles. Airports and railroad stations must have space for long-term parking, Fig. 11-21.
5. Store cargo. Warehouses are built next to docks and airports. Goods are kept there temporarily.

A
B
C

Fig. 11-19. Airports must have several types of on-site transport. A—People enter and leave aircraft on portable steps or ramps. B—A moving sidewalk is a type of "people mover." C—Luggage is moved with conveyors, and baggage carts.

Fig. 11-20. A plane is serviced and refueled at the boarding gate.

Fig. 11-21. Vehicle storage is important. Travelers need space to park automobiles.

HOW TRANSPORTATION DEVELOPED

It took thousands of years to develop transportation systems. People had modes of travel before there were written records. We are not even sure when the first transportation system was developed.

Our ancestors (earliest relatives) had to search for food. It sometimes meant long trips in search of wild game, nuts, and berries.

As they learned to use tools, drags and sleds were made. Homemade versions of the drag were used by American pioneers, Fig. 11-22.

However, three developments greatly affected the future of transportation. They were important advances in their own time. They are used in all modern transportation systems. Where would we be today without:
1. The wheel.
2. The heat engine.
3. The discovery and refining of petroleum.

The Wheel

Just about every movable item you can think of has wheels. Many toys have wheels. Any kind of machine that travels has them.

People used no wheels until around 3500 B.C. We do not know how they developed. Perhaps the first ones were simply logs. The logs were placed under heavy loads. When the load was pulled, the logs rolled. Later, someone attached a small log to each end of an axle. Crude ropes secured the axle to the load, Fig. 11-23.

Invention of the Heat Engine

Wheeled carts pulled by animals meant that people could travel must faster. They could also move more goods faster. But even the strongest animals tire.

Fig. 11-22. Sled or "stone boat" used by early American farmers was a refinement of primitive drags.

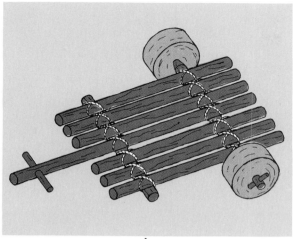

A

B

Fig. 11-23. A—Artist's idea of early wheeled vehicle. Wheels were sections of logs. B—The covered wagon was important to settlement of the American West.
(U.S. Park Service, Natchez Trace Parkway)

The invention of the heat engine was an important advance. The engine could work faster than animals. It did not get tired.

The first engines ran on steam. Water was heated in a closed container. Steam was allowed to escape to another chamber. Here it moved a piston. See Fig. 11-24.

On the first steam engines the piston was hooked to a rocking arm so the engine would pump water. Then other inventors connected the piston to a crankshaft and a wheel. Then it could move vehicles.

Gasoline and diesel engines

Another type of heat engine was developed. It burned fuel inside the engine (internal combustion). It was smaller than the steam engine. Also, it did not need to carry as much fuel.

The internal combustion engine is designed to draw in fuel and air, compress it in a closed chamber, ignite it to produce power, and then expel the spent fuel charge. The heat engine replaced animals for use in transportation.

Henry Ford made automobiles and the gasoline engine popular. He found a way to mass produce them (make them in large numbers). He could sell them cheaply, Fig. 11-25.

Early humans also made simple vessels to travel on water. First, logs were used. Then the logs were hollowed out to make dugouts, Fig. 11-26. Indians made canoes. They attached birch bark to a crude frame. Eskimos stretched skins over a wood frame to make a kayak. Boats made of bound reeds took Polynesian sailors on long ocean voyages.

The Phoenicians, Greeks, and Romans traveled on the Mediterranean Sea. Their ships

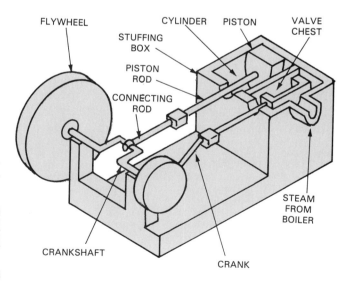

Fig. 11-24. Simple drawing of early steam engine.

Fig. 11-25. The first assembly line for manufacturer of automobiles was set up by Henry Ford. The year was 1914. The line was outdoors! (American Petroleum Institute)

Fig. 11-26. Reproduction of a dugout canoe. Ancient boat builders hollowed out logs using axes or fire. Dugouts can still be seen in the Pacific islands.

had to be rowed. To get more power for larger cargo, they had banks of oars and oarsmen.

Later, sails were added to use the power of the wind. The invention of the sail made it possible to carry large loads to distance places. Only a small crew was needed. Most of the space could be used for cargo. This change affected the whole world. It became possible for people to travel farther. They found new land. They could trade goods with people all over the world. Look at Fig. 11-27.

Steam engines were originally built to propel land vehicles. An American, Robert Fulton, developed the first successful steamboat. Two steam-powered ships crossed the Atlantic in 1838. Within the next 40 years sailing ships were almost completely replaced by steam vessels.

Developing air transportation

For thousands of years, people only dreamed of flying, Fig. 11-28. From time to time someone would try to imitate birds by attaching wings to their arms. However, the first flights by humans were not in winged vehicles.

The Montgolfier brothers in France developed lighter-than-air balloons large enough to carry people. On June 5, 1783, they launched a balloon made of linen.

Fig. 11-27. Invention of the sail made long ocean voyages possible. (American Petroleum Institute)

Fig. 11-28. Watching birds in flight, humans dreamed of flying.

Fig. 11-29. The Wright brothers, Wilbur and Orville, interested in flying from boyhood, built and flew the first successful powered airplane. Their first flights were made in 1903 at Kitty Hawk, North Carolina. The plane had a 12 horsepower engine designed by the Wrights. The plane was launched along a wooden rail (left).   (Smithsonian Institution)

Leonardo da Vinci thought and studied a great deal about vehicles that could fly. His work on designs for aircraft lay forgotten or unknown until the 19th century. By then, aircraft were commonplace (many were in use).

Three devices important to aviation were already developed in da Vinci's time:

1. The windmill. It was probably the earliest model for the propeller. Ancient Persians used the windmill to capture the wind's energy.
2. The kite. Like the airplane wing developed much later, it caught the lifting action of rushing air.
3. The model helicopter. This was a toy known in the 15th century. It has been popular ever since.

However, the Wright brothers of Dayton, Ohio, built and flew the first practical airplane in 1905. See Fig. 11-29. It could bank, turn, circle, and make figure eights. It was able to remain in the air for half an hour at a time.

INPUTS FOR TRANSPORTATION

Like other technologies, transportation needs inputs. Inputs are the resources invested (spent) on an activity.

The inputs for transportation are very much the same as those needed for other technological systems. They are sometimes called resources. It is the way the resources are used that makes one system different from another.

Transportation inputs include:
1. People.
2. Energy (such as fuel).
3. Information.
4. Tools, materials, and mechanisms, (machines).
5. Capital and finance.
6. Time.

People as a resource

People provide the intelligence. They design and organize transportation systems. Drivers operate vehicles over roads and highways. Pilots fly airplanes and control ferryboats and ships. Navigators keep track of a ship's progress across oceans, Fig. 11-30.

Others provide training, plan trips, sell tickets, or handle baggage and cargo, Fig. 11-31. Maintaining equipment is very important. Mechanics service and repair engines and other mechanical systems. Maintenance people clean and paint. Still others look after passenger's comfort and wellbeing. Food and supplies must always be available.

Information resources

Many people must have special knowledge to make transportation systems run smoothly. There are many specialists. We talked about some of them in people resources. Managers must be trained to organize their departments. Ticket agents must have information about schedules, prices, and travel times. Drivers, pilots, and those who control ships and trains, need a continuous flow of information to

Fig. 11-30. Third mate, Susan Janis, is the navigator on an oil tanker. (Exxon Corp.)

Fig. 11-32. Airline passengers read schedules to get information on flight times.

Fig. 11-31. Training or personnel is one responsibility of management. This instructor is operating a flight simulator. It is used in training airline pilots. (Harris Corp.)

Fig. 11-33. Railroad personnel are engaged in routing trains. They use computers to receive, process, and send information. (CSX Corp.)

operate their vehicles, Fig. 11-33. They need to know weather conditions, speeds, conditions of systems in their vehicles, routes to travel, and safety information. Many tools and instruments are needed to provide this information. Among them are:

1. Two-way radios.
2. Radar.
3. Road signs.
4. Navigational instruments.
5. Gauges, speed indicators, mileage indicators, and on-board computers.
6. Maps.

Tools and mechanisms

Tools are needed to keep vehicles running. Engines must be overhauled. Tires must be changed. Bodies must be painted. Damaged parts must be fixed or replaced. Modern automobiles and planes need to be maintained with analyzers and automatic test equipment.

Materials

Materials such as rubber, steel, iron, plastic, and aluminum are used to build the machines used in transportation. Fig. 11-34 shows a factory where aircraft are assembled.

Transportation companies have offices where records are kept. Payrolls must be maintained. Typewriters, computers, and business machines are used.

Fig. 11-34. Aircraft are assembled from many different materials. The fusilage (body) is lightweight materials, usually aluminum. (BSU)

Time

It is important for people and materials to move from one place to another on time. For this reason, time is a resource that must be scheduled. Printed schedules inform shippers and travelers when buses, trains, or planes leave and when they are due to arrive at their destination. People regulate their travel activities by using the schedules, Fig. 11-35.

VEHICLES AND THEIR SYSTEMS

Transportation technology is built around the vehicle. This vehicle must be designed to suit the purpose. It must be able to transport

Fig. 11-35. Airline passengers get travel information by checking airline television screens.

using one of the modes described earlier. Every vehicle must have:

1. A structure. Vehicles are movable structures. They hold not only people and goods but the means of moving the structure from one place to another along a path. The structure also provides a rigid framework to support other systems.
2. A means of propulsion. This means a power source to move the vehicle. Usually, this is some kind of engine.
3. A means of transmission. Power must move from the propulsion unit to where it will cause the vehicle to move.
4. A guidance system. This system receives information.
5. Control systems. These enable the vehicle to change speed and direction.
6. Measurement devices which monitor (check) that the vehicle is operating properly.

Vehicular Structures

You are accustomed to riding vehicles. Maybe you had a tricycle when you were a small child. It had all the necessary parts of a vehicle. It was a complete system. The tricycle had a frame. The frame supported all the other parts. The frame was the structure of the tricycle.

Like the tricycle, any transportation vehicle has a framework that supports all of the vehicle's parts, Fig. 11-36. The framework has a covering.

Propulsion Systems

Heat engines are the most common form of power for moving vehicles. We know them as:
1. Gasoline engines.
2. Diesel engines.
3. Jet engines.
4. Gas turbines.
5. Rocket engines.
Do you remember learning about them in Chapter 4? Fig. 11-37 shows a typical car engine.

Transmission of Power

A power source, or propulsion unit must have a way to move power to do work. This is called transmission. It means to move from one spot to another. Power can be transmitted by:
1. Belts or chains. (A bicycle chain moves power from the bicycle pedals to the rear wheel.)
2. Gears. These are notched wheels that mesh (fit together). One gear will drive another.

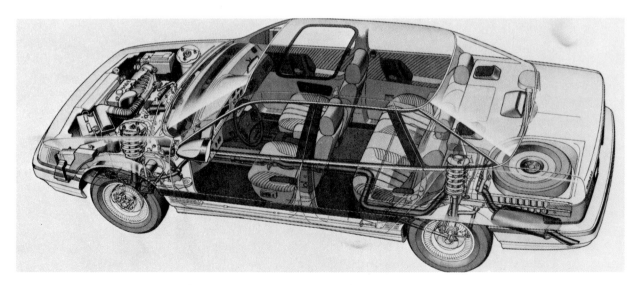

Fig. 11-36. Phantom view of an automobile shows all of its systems.

Fig. 11-37. Cutaway shows typical engine for an automobile. (Buick)

3. Shafts. (A shaft carries an engine's turning action to the drive wheels.)
4. Fluids. (Compressed air or liquid can lift or move objects. For example, pressure on the water makes water fountains work in your school.)
5. Electricity or electromechanical means. (Electricity can be transmitted through a conductor such as a wire. The electricity can then drive a motor. The motor causes the force or motion.)

A robot is an example of a mechanism that can be controlled and moved by several methods. It may use electromechanical devices. It may have gears and shafts. Chains may be used to make some movements.

To transmit its power to move a vehicle, the propulsion unit must be attached to the vehicle. The jet engine moves forward from the greater pressure of gases on the forward end of the engine. This force is transferred through the engine. Because the engine is attached to the frame of the aircraft, the aircraft moves too. However, the motion of many heat engines

is rotary (circular). Another mechanism called a transmission must transfer the rotary motion to the wheels.

In the simplest transmissions, two discs are used. One is attached to the crankshaft of the engine. It spins with the crankshaft. The other is attached to the drive shaft and drive wheels. When power is needed to drive the vehicle, springs press the two discs together. The one attached to the engine is already spinning. It causes the other to spin also.

Gears at the other end of the drive shaft transmit the power to the wheels. They are connected to the wheels by long shafts.

Guidance Systems

Guidance systems are not part of the vehicle. They provide information to the vehicle's operator. Aircraft receive instructions from a flight controller. This person directs the aircraft into and away from the airport. The controller tells the pilot when to take off or land, what course to follow, and what height to fly at.

Railroads have signals along the right-of-way. Highways traffic gets information from stop signs, stop lights, and other roadside signs. Fig. 11-38 shows a typical guidance system on a waterway.

Control

The operator must be able to stop, start, speed up, slow down, and turn the vehicle. This is arranged through systems of control. They vary somewhat from vehicle to vehicle.
1. Braking system. Trains and highway vehicles and airplanes have wheel brakes. Air or hydraulic pressure pushes pads or blocks against the wheels. Friction produces drag on the wheels to slow or stop the vehicle. Airplanes also have wing flaps. They produce air drag and help slow the plane. Ships are slowed or stopped by reversing the propellers. Aircraft engines are also used to slow the plane after landing.
2. The amount of power needed to move a vehicle will vary. The operator must be able to control the amount of power. Acceleration and deceleration controls will vary

Fig. 11-38. Towers in foreground hold radar equipmente and television cameras. They help control vessels on the canal. (Eaton Corp.)

speed by controlling the amount of fuel delivered to the engine. Fig. 11-39 shows controls for brakes and acceleration.
3. Vehicles also must have directional control. Steerable land vehicles have wheels that turn left and right, Fig. 11-40. Ships and airplanes have rudders for left and right movement. Airplanes also control up and down movement with an elevator. It is part of the tail assembly.
4. Miscellaneous controls. These include switches to turn on lights, windshield wipers, windshield washers, defrosters, heaters, and radios. These are usually simply "on-off" electrical switches. They control electrical current to the devices.

Measurement Devices

The driver must receive information about the operation of the vehicle's systems. This information is provided by dials and gauges. Look at Fig. 11-41.
1. A fuel gauge tells the operator how much fuel is left.
2. Electrical gauges indicate whether the vehicle's alternator is generating enough electricity. It is important that the battery does not discharge. A discharged battery is not able to start the vehicle.
3. Speed indicators measure how fast a vehicle is traveling. An odometer measures distance the vehicle has traveled. It alerts the driver to have the vehicle serviced. It also gives information about distance traveled during a trip.
4. Temperature gauges register the temperature of the engine coolant. They alert the driver of overheating problems so proper repairs or service can be performed before the engine is damaged.

Computer Control

Modern engines are controlled by a small computer, Fig. 11-42. It is carried in the vehi-

Fig. 11-39. Brake and gas pedals are part of this control system for an automobile or truck. (American Petroleum Institute)

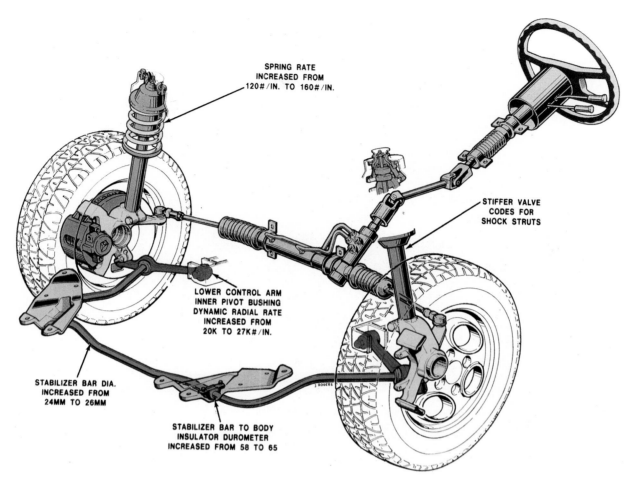

Fig. 11-40. View of steering control for an auto. It gives the operator directional control so he or she can steer it.   (Ford Motor Co.)

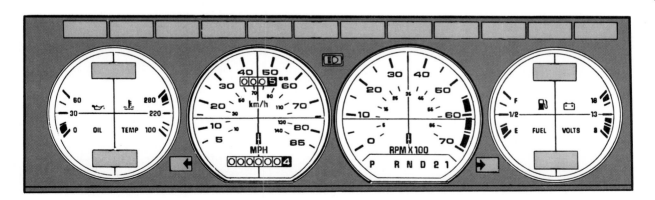

Fig. 11-41. Gauges tell operators the operating condition of a transportation vehicle. They measure speed, fuel, electrical system and check various parts of the system.

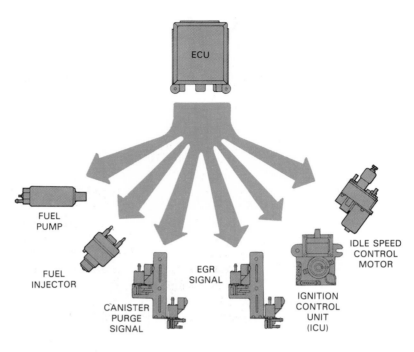

Fig. 11-42. The computer (marked "ECU") sends signals that control the operation of many different systems of an engine. (Renault)

cle. It is programmed (set) to control different operations such as:

1. Fuel mixture for best operation of the engine.
2. Timing (when spark is delivered to each cylinder).
3. Shifting of automatic transmissions.

How it works

The computer has a memory which tells how the vehicle engine is supposed to work. It receives messages from **sensors** located where they can "tell" how the engine is operating. The sensor reports engine conditions as electrical signals. The computer "knows" what the signals mean. It uses this information to adjust the operation.

For example:

1. If the exhaust sensor signals that the engine is getting too rich a fuel mixture. The computer will reduce the amount of fuel in the air-fuel mixture.

2. If the sensor signals that the engine is "pinging" because the spark is delivered too late, the computer will advance the spark.

The computer can make these changes by its control of **actuators.** They are devices which use an electric current to cause movement. One type is the solenoid. It is somewhat like a magnet. Solenoids have a back and forth motion. They can open and close switches or valves. Small electric motors are another type of actuator. They adjust engine systems by turning parts to adjust them.

SUMMING UP

Transportation is the moving of people and goods from one place to another. Modern transportation includes several modes: land, water, air, and space. When more than one system is employed it is called intermodal. Every mode of transportation has support systems. These are the terminals, docks, and repair shops.

Early humans had to walk and carry their goods until they began to develop crude vehicles. Eventually they built vehicles that could travel on land, and ships to sail the seas. Today transportation is important to our lives. Air, land, and sea systems transport food and goods. They also allow us to travel and visit faraway countries

Like other technologies, transportation has inputs, processes, and outputs. Inputs are resources such as people, tools, materials, and machines, information, energy, capital, and time.

Processes are what we do to make transportation work. One process is management. These are the planning, organizing and controlling or supervising activities. Production processes are the actions that do the moving. It includes loading freight, selling tickets, seating passengers, driving, piloting, or steering a vehicle or ship.

A transportation system is made up of structures and pathways. The structures are buildings, and vehicles. Pathways are roadways, airlanes, and waterways. Pipelines and escalators are examples of other kinds of paths.

Vehicles are made up of several subsystems. The subsystems are supported by a frame which contains the people or freight. The subsystems include: a means of propulsion (such as an engine), a transmission, controls, and measuring devices.

Computers are a recent addition to the control subsystem of a vehicle. They have taken over many of the tasks once done by the driver.

KEY WORDS

There words were used in this chapter. Do you know their meaning?

Actuators, Cargo, Conveyors, Ferries, Four-cycle engine, Freighters, Heat engine, Helicopters, Intermodal transportation, Ocean liners, Pipelines, Rockets, Rolling stock, Satellite, Sea lanes, Sensors, Shipping lanes, Tankers, Tourism, Transportation, Tugboats, Unit train.

TEST YOUR KNOWLEDGE
Chapter 11

1. Transportation is (select correct answer):
   a. Airplanes, automobiles, and ships.
   b. Carrying people from one place to another.
   c. Moving people and goods from one place to another.
   d. Vehicles.
2. List the four modes of transportation and define each.
3. What do a highway, railroad track, a shipping lane and an airway have in common?
4. What does the term "degree of freedom" mean when we talk about modes of travel?
5. A _____ train hauls only one type of cargo and to one place only.
6. List the main advantage of a helicopter over an airplane.
7. Space craft must carry their own oxygen so the rocket engines can burn their fuel. True or False?
8. What are support systems for transportation and why are they needed?
9. What three developments most affected the future of transportation?
10. Explain how the wheel might have developed from ancient times.
11. The first heat engines were:
    a. Gasoline engines.
    b. Steam engines.
    c. Jet engines.
12. _____ are the resources invested in an activity.
13. Indicate which of the following are not transportation resources:
    a. Driving a vehicle.
    b. Time.
    c. A tankful of gasoline.
    d. A set of mechanics tools.
    e. A schedule for a commuter train.
14. A plane's flying time between two cities is a _____ resource.
15. When speaking of transportation vehicles, what does the term "structure" mean?
16. Name three operations that a computer might control on a car.

Busy highways and expressways have become part of our lives. Legislators, city officials, and traffic engineers are looking for ways to relieve the traffic congestion (jams) around heavily populated areas. What do you think could be done?

Railroads were important in the settling of the United States and Canada. This is a steam locomotive of the late 1800s. It is preserved at a railroad museum in Kentucky.

On-site transportation. This heavy-duty dump truck transports rock from a quarry area to a crusher on the site.   (Caterpillar Inc.)

APPLYING YOUR KNOWLEDGE

Introduction

Every day, millions of people and countless tons of cargo are moved. They are transported from one place to another. Technological systems are used to move these goods and people. Energy is used by vehicles to haul cargo. People feel power applied to engines that move these vehicles over land, across water, and through the air.

These transportation systems include vehicles and pathways. The pathways allow the vehicles to move from one place to another. The pathway may be a highway or railroad track. Or it may be a river or an ocean. Air and space provide pathways for airplanes and spacecraft.

Each of these systems is designed by people. The systems allow us to extend our potential for movement. This activity will allow you to design and test a vehicle for a transportation system. You will use your creative ability and manual skill to design and test a boat hull.

Equipment and Supplies

Waterway or long narrow channel that holds water.
2" x 4" x 9" styrofoam block (hull)
2 1/2" x 3 3/4" x 28 gauge sheet metal (keel)
Small eye screw
Sail material:
   6" x 6" square of fabric
   2 Dowels 3/16" x 6"
   1 Dowel 3/8" x 7 3/4"
Rule and square
Scratch awl
Tin snips
Mill file
Needlenose pliers
Tailor's chalk or fine-line magic marker
Fabric glue or needle and thread
Hot wire styrofoam cutter or bandsaw
Coping saw

Rasp or "surform"
Coarse abrasive paper
Compressed air (15 psi) or a fan and a cardboard tube
Stop watch or a timepiece which reads in seconds.

SAFETY NOTE: The tools and equipment needed for this activity can cause injuries if not handled properly. Your instructor will provide safety rules. He or she will also demonstrate safe procedures. Do not use any tool or piece of equipment unless you know how to do so with safety.

Procedure:

Making Hulls:
1. Your teacher will divide you into groups. Each group should have three students.
2. Each group will build three basic hulls as shown in Fig. 11A.
   a. Student 1 will build a round bottom hull.
   b. Student 2 will build a V bottom hull.
   c. Student 3 will build a flat bottom hull.
3. Carefully watch your teacher demonstrate safe techniques for cutting and shaping styrofoam.
4. Each group should get three styrofoam blocks.

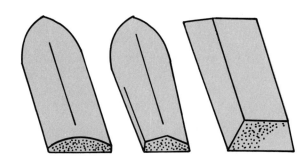

ROUND BOTTOM   V BOTTOM   FLAT BOTTOM

Fig. 11A. Boat hull designs. Each group will make one of each design.

5. Each member of the group will:
   a. Design a hull that meets his or her assigned shape.
   b. Cut the styrofoam to the basic shape using a hot wire cutter or bandsaw.
   c. Smooth the hull shape using a rasp, "surform," and/or abrasive paper. Be sure that both sides of the centerline are alike.

Making Sails:
1. Select the materials listed under equipment and supplies for each sail.
2. Layout the fabric using tailor's chalk or a fine-line magic marker. See Fig. 11B.
3. Fold the fabric at each end to form a sleeve.
4. Sew or glue the fabric in place.
5. Cut two pieces of 3/16 in. dowels, 6 in. long, using a coping saw.
6. Insert the dowels in the top and bottom sleeve of the sail.
7. Cut a piece of 3/8 in. dowel, 7 3/4 in. long using a coping saw.
8. Shape the dowel, as shown in Fig. 11C, to make the mast.
9. Insert mast in the sail as shown in Fig. 11D.

Making a Keel:
1. Select a piece of 28 gauge sheet metal.
2. Lay out the keel as shown in Fig. 11E.
3. Cut out the keel using a pair of tin snips.
4. File all edges with a mill file to remove burrs.

Assembling the Boat:
1. Insert the keel into the center of the bottom of the boat.
2. Attach the sail to the top of the boat.
3. Slightly open the eye of the eye screw with a pair of pliers.
4. Install a eye screw in one side of the boat. This will attach to a string in the waterway. The string will keep the boat traveling in a straight line in the waterway.
   NOTE: Each group should now have three boats: a square bottom, a round bottom, and a V bottom model. Each boat should have a keel and a sail in place. They should also have a eye screw in one side.

Testing the Boats:
1. Your teacher will provide the class with a waterway to test the boats. The waterway will have a string or fish line stretched along its length.

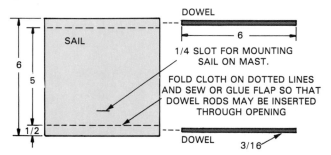

Fig. 11B. Make the sail from cloth and dowels.

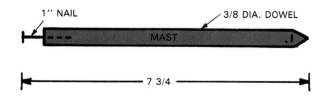

Fig. 11C. The mast should be tapered at one end for insertion into the styrofoam hull.

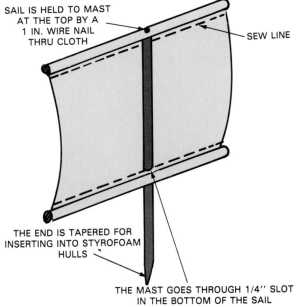

Fig. 11D. Assemble the sail as shown.

2. Place the boat at one end of the waterway.
3. Hook the screw eye over the guideline.
4. Give the sail a continuous blast of 15 psi compressed air or use a fan to blow air through a cardboard tube.
5. Clock the time it takes the boat to reach the end of the waterway.
6. Record the hull type of the boat.
7. Record the time it took the boat to reach the other end.
8. Determine the average travel time for:
   a. The flat bottom boats.
   b. The round bottom boats.
   c. The V bottom boats.
9. Compare the results and determine which shape is most efficient.

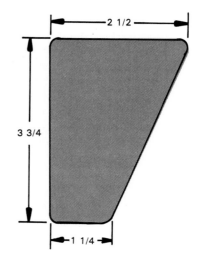

Fig. IIE. Shape the keel out of sheet metal.

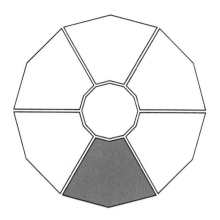

Technology and Society

Our technology is very much a part of our lives. The more we control and change our environment the more it affects us. It affects how we live. It affects how we will act.

Winston Churchill was a famous prime minister of England. He once said, "We shape our buildings and then they shape us." He was expressing the same idea.

Can you understand what he means? Let's consider how technology affects our values. Suppose that you had to live in a tent because your community had to move often to make a living. Would you value the same possessions you and your family have now? Probably not. Consider the things you have in your bedroom. How many of them would you want to keep if you were moving every month? Would you prefer a bed? Or a sleeping bag? Would you want a chest of drawers, or a duffel bag?

Do you see how technology has become a part of us? We may live all of our lives in one community. If we do move, technology provides large moving vans so we can take bulky furniture with us. There are many other examples of how technology has become interwoven with our day-to-day living. You have read about these technologies in preceding chapters.

This is not to say that all of the impacts (effects) of technology have been good. We have to decide what is best for the future. At the same time we need to look ahead and see what new technologies might be coming.

In the following two chapters you will have a chance to discuss the impacts of technology. You will learn how society is now looking at the effects of technology on environment. You will learn how "Futurists" make predictions about technological advances we don't even dream about now.

We are learning more about what is good about our technology. We are also learning what we must not do to avoid damage to our environment. It is important for us to know these things. For we cannot go back to the primitive life that our ancestors knew.

Technology is not bad in itself. However, by not using it wisely we can cause pollution and erosion. Better farming methods will correct the washing away of soil. (USDA Forest Service)

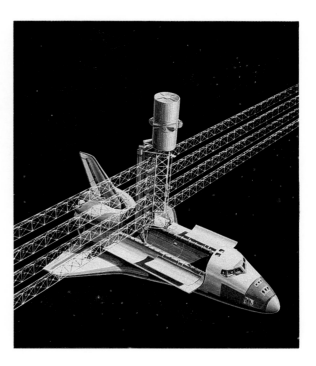

Space manufacturing. This beam builder could one day make parts of a space solar power system. The beam factory, itself, would be floating in space. (NASA)

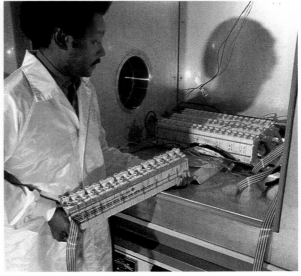

Who can truly say what technology is in our future? Will we be building in space? Perhaps. Will we all be riding in solar-powered cars? Possible. Left. A solar-powered car designed to travel at speeds up to 45 mph. Right. Solar power is stored in a battery with 68 silver zinc cells. (GM Hughes Electronics)

Chapter 12
Impacts of Technology

The information given in this chapter will help you to:
- ☐ Explain how technology has made life better for humankind.
- ☐ Discuss some undesirable effects of technology relating to pollution, jobs, and resources.
- ☐ Suggest how some of the consequences of technology can be avoided.

The impact of technology means its effects. Technology has an impact on each one of us.

It also has an effect on our surroundings. If by riding a bicycle, you can save five minutes getting to school, the bicycle has a good effect or impact on your life.

But what of the bicycle's effect on the surroundings? The bicycle must have a smooth path with a hard surface. Providing the path takes land. The land may not be used for other purposes. Then the bicycle must have a storage place at home. It must also have a space at school too. The storage may be a building or a simple bike rack. Both have some effect on the surroundings. They take up space, Fig. 12-1.

Fig. 12-1. Transportation vehicles get us places faster. But when we get there we must park them. Bicycles take up less space than cars.

An automobile will get you to school faster than the bicycle. It will also shelter you from heat or cold, rain and snow. However, its effect on the surroundings are much greater than the bicycle. The path must be widened into a street or highway, Fig. 12-2. The parking space at home or at school takes up more space. A garage will also take up more space than bicycle storage.

There are other effects from the automobile as well as from the other products which we use. We will talk about these later.

Some of the effects of technology are good. Others are not. Some of the effects have been intended. Others have been an unhappy surprise.

Fig. 12-3. Technology gives us better food.

Fig. 12-2. Land must be set aside for roadways and paths for vehicles.

Technology Produces Change

An important thing to realize about technology is that it produces change. Most of the things you use or eat are the result of technology. Humans use tools, materials, and knowledge to make these changes.

Why do we change the way we do things? Usually there are two reasons:
1. We want to improve our environment. We want our country, our city, and our home to be better places to live.
2. We want more time to enjoy ourselves. This extra time is called **leisure.**

Technology has given us these good things. Because of technology we have better food, better houses, and better means of moving about, Fig. 12-3. Because of technology, we have better ways of keeping in touch with each other. Our health is better because we have better **nutrition** (diets). Better medicines can cure us when we are ill.

How Technology Has Caused Change

All of technology is linked together. Better buildings and goods became possible because technology developed new materials. The knowlege that helped develop the new materials had to be spread to many workers. This meant that methods of communicating had to change. Technologies improved. Resources improved. This gave us improved materials.

Changing materials and resources

At one time there were only natural materials. Tools were made of sticks, stones, and plant fibers. Before drilling tools were developed, only energy found above the ground or in the water could be used. Wood, coal, and animal oils were the only fuels known. They were the only means of producing heat, light, and power. When iron and steel were developed, drills could be made. The drills could bore deep into the earth, Fig. 12-4.

Fig. 12-4. Drilling for natural gas in Alberta, Canada. Air and suds from a detergent help bring cuttings to the surface.
(Standard Oil Co., Indiana)

Petroleum and gas were found. As fuels and raw materials, they have made possible the great advancements in all technologies.

We have materials that our ancestors never knew. Technology has taken natural materials and changed them so they have new properties. For example, iron can be treated to make it less brittle and stronger. Thus, it can hold heavier weights. Plastics, Fig. 12-5, are a rather

A

B

Fig. 12-5. Many products today are made of plastic, a new material derived (gotten) from petroleum. A—Boat hulls of fiberglass need less maintenance. B—Countertops are made of plastic laminate and dishes are made of a durable plastic. (Marshfield Homes)

new material that have been developed from petroleum. They are strong and neither rust nor corrode. Many things that were made of metal are now made of tough plastic. New wrinkle-free, easy care clothing are made of fibers developed from plastic. Even damaged human body parts can be replaced by plastic. Artificial limbs, joints, and hearts are built of a tough, long-wearing plastic.

Designing new tools and machines

Have you ever visited a museum? You may have looked at the old tools shown there. You could have seen how different they are from the tools we use today.

If you look at the products produced by these old tools two things will occur to you:
1. Today's tools do not require as much labor.
2. The products, such as furniture and clothing, produced with the older tools were much cruder.

Also, at museums you can study the structures built hundreds of years ago. They were often built on a smaller scale than buildings of today. Why do you suppose that is?

Isn't it because most of the work had to be done by hand? In those times, the timbers in a house had to be shaped with axes or hand-saws. Today, power saws and planes do the shaping quickly.

Yes, new materials allowed larger buildings. But tall buildings became possible with the advent (coming) of steel columns and beams. This meant people could make better use of the land. See Fig. 12-6. More people could live and work in less space. This has made the building less expensive and more useful.

Keeping in Touch

A popular commercial in 1987 invited us to "reach out and touch someone." It encouraged people to use the telephone to keep in touch with family and friends who were far away. The phone is one of the communication devices found in most homes. It is possible to dial or punch in a series of numbers and speak to people anywhere in the country, Fig. 12-7. It is not necessary to have an operator put through the

Fig. 12-6. Having better tools means we can build better buildings.
(Prestressed Concrete Institute)

Fig. 12-7. We keep in touch now by using the telephone.

calls. Technology has developed automatic devices which make the proper connections from the code numbers you have used.

Other devices have made it easier for us to keep in touch with our modern world. News of what is happening anywhere is flashed to us in an instant. Signals "bouncing" off satellites orbiting in space beam sound and pictures into our homes. We are entertained by radio and television.

Perhaps we take these communication devices for granted. They have been with many of us all our lives. Our ancestors, on the other hand, never had them. Alexander Graham Bell invented the first crude telephone in 1875, Marconi the radio in 1900. The first radio station went on the air in 1920. About 25 years later television sets began to appear in homes. Ask your grandparents or an older person how television affected their lives. What did they do for entertainment before that?

At one time calculators were considered the latest and best device for making calculations. Now computers work problems faster. They are also taking the place of typewriters for processing correspondence and producing manuscript for books.

Getting Away

Before 1900 "getting away" probably meant hitching a horse to a buggy and taking a drive in the country. Now it may mean a trip to the airport in a bus or automobile. Then a plane can whisk you across the country in a matter of hours. After thousands of years of development, transportation quickly went from the steam engine to the jet engine in little more than 100 years.

In the time it once took to travel 1000 miles, rockets have carried astronauts to the moon and back to earth. Foods can be picked, packed, flown 2000 miles, and eaten the same day. Oil pumped in the Near East can be refined in Indiana. Later it is used to drive a vehicle in California. Indeed, transportation and distribution of goods and foods has changed drastically. It truly makes our life easier and more enjoyable.

With technology, it is possible to be more **productive.** Being productive means doing work that contributes something good to life. If you clean up your room you have done something that makes your surroundings more attractive. Your friends may be more comfortable visiting you in an orderly room.

We like to be productive for several reasons:
1. If we work at a job, we produce goods or services that benefit others. They may benefit the whole community or just certain people. A farmer who raises chickens and sends eggs to market benefits everyone. He or she provides necessary food. A toymaker who makes dolls benefits children. Both persons are contributing something good.
2. We need work to earn a living. In our society everyone is paid for the work they do. This is no more than fair. We produce something that is needed. We are paid for the service. In turn, we can purchase services and products that others produce.
3. Another reason for being productive is the good feeling it gives us. People want to feel that they are useful. One complaint of old persons is that they no longer feel needed.

Technology has provided work for people. Remember earlier chapters where we talked about what it was like to live in primitive communities? Most of their time was spend hunting for food. Some people were good at gathering food; others were not. Not everyone could feel useful.

Technology changed that. As technology grew, so did the variety of things that people could do for a living. Farming tools were invented. Then people who liked to grow things could farm. When tools for construction were developed some people became builders. As you can see, technology provided a variety of jobs. It still does today. We can see that technology has been good about providing work. See Fig. 12-8.

NEGATIVE IMPACTS

We have shown how technology has had a good impact on human life. But remember

A

B

Fig. 12-8. Even 30 years ago farming required more workers. A—Smaller tractors were used. More hand labor was needed to harvest crops. B—Today's harvest combines are larger and more powerful. One or two operators do the work of many farm workers.   (Gulf Oil Corp.)

what we said at the beginning of the chapter. Not all of the effects of technology are good.

This is not because technology is bad. Rather it is because we have not always understood technology. We haven't always used it wisely.

Scarcities

One negative impact of technology is that its unwise use often creates **scarcity.** The steel, plastic, and woods used to make our life easier come from raw materials. But these are not available in unlimited quantities. The earth contains only so much of these materials. If we use them up they will be gone forever.

An earlier chapter talked about energy needs. Energy produces power for our technological devices. We depend heavily on oil. Again, nature controls the supply. When it is gone there will be no more.

The impact of scarcity drives humankind to do three things:
1. Save the nonrenewable resources we have left by using them more wisely than in the past, Fig. 12-9.
2. Develop and use renewable resources.
3. **Recycle** used materials.

Fig. 12-9. Shortages of fuel affect each of us. How would a shortage of home heating fuel affect you? Or your use of an automobile? (Sun Oil Company, Inc.)

For example, we now import about 50 percent of our petroleum from foreign countries. About a third of all oil consumption is for transportation. Much of the fuel is burned in automobiles. Home heating also requires oil and natural gas. See Fig. 12-10.

People are being forced to make choices on how to use the petroleum and natural gas fuels wisely. Automobiles are now designed to use less fuel through:

1. Smaller and more efficient engines.
2. Lighter bodies.

New methods of building homes are making them easier and cheaper to heat. Builders are putting more insulation in walls and ceilings, Fig. 12-11. They are installing windows with two and three panes of glazing (glass) in them. The air space between panes of glass does not transfer as much heat. By sealing up cracks during construction builders also prevent loss of heat by convection (air passing through). See Fig. 12-12. Some new homes are being designed to use solar heating to save fuel. Perhaps as you've traveled you've seen homes built into the side of a hill or almost entirely underground. Sometimes **earth berming** (piling earth along one or more sides) is used to save energy. Refer to Fig. 12-13.

Fig. 12-10. The average American or Canadian family uses about 180 barrels of petroleum a year. This is enough to fill the 138 55-gallon drums shown here. (American Petroleum Institute)

Fig. 12-11. Wall is being doubled up so more insulation can be placed between the studs (vertical lumber). (Fiberglass Corp.)

A B

Fig. 12-12. Modern homes are being made so that less fossil fuel is burned to keep them warm. A—Exteriors are "wrapped" in a tough plastic film to seal cracks that let heated air leak out. B—Solar panels in the roof help heat this building.

Fig. 12-13. An earth sheltered house has earth over or around it. This makes it easier to heat.

Damage to Environment

Not all of the changes technology makes have been good for the environment or people. Technological activities which have improved how we live have also caused problems for some people and damaged our surroundings. This damage includes:
1. Air pollution.
2. Water pollution.
3. Noise pollution.
4. Erosion, **depletion,** and other damage to soil.
5. Poor working conditions and dislocations of workers.

Air pollution

Air pollution is putting dust, fumes, smoke, gases, and other material into the air. Each year air pollution costs millions of dollars in damage. Crops develop disease or blight. Buildings are damaged and metals corrode. If heavy enough, the pollution can be harmful or fatal to humans.

Some pollution is caused by nature. Plant pollen and dust are stirred up by winds. Pine trees contaminate the air with **terpenes.** (Terpenes are a hydrogen and carbon material. The trees release them as a gas.) However, the pollution caused by human technological activities is more serious when it is released to the air in great quantities.

One of the greatest sources of air pollution is the burning of fossil fuels, Fig. 12-14. This releases unburned carbons into the air. Other sources are harmful chemicals that can be introduced into the air by crop dusting and industrial accidents which occur during manufacture of chemicals. A large scale release of chemical fumes, for example, can be deadly to all life. In 1984, the accidental release of a toxic gas at a pesticide (chemicals used to kill insects) plant in India killed more than 2300 people. Another 30 to 40 thousand people were seriously injured.

The by-product of a nuclear power plant is highly radioactive wastes. Release of steam containing radioactive material is always a danger. Winds carry the waste over a large area.

Several accidents have occured in nuclear power plants. In the 1970s, radioactive steam was released accidentally at Three Mile Island nuclear power station in Pennsylvania. In 1986, a nuclear power plant blew up at Chernobyl, Russia. It contaminated a large area of Russia.

A

B

Fig. 12-14. Burning of fossil fuels had become a serious problem by the 1960s. Now environmental laws are dealing with it. A—Factory smoke stacks from years ago. "Scrubbers" and other technologies take many of the pollutants out of burning processes. B—Auto exhaust gases caused smog problems in Los Angeles. (American Petroleum Institute)

Water pollution

Disposal of human wastes, growth of manufacturing, and farming practices have also caused water pollution. The most common contaminants (waste materials) are industrial wastes, sewage (human wastes), and agricultural runoff containing fertilizers, weed killers, insecticides (chemicals that kill bugs) or minerals leached out of the soil. (In leaching, pollutants in the soil dissolve in the water.) Another material that pollutes is laundry detergent. Also, acid in the air is collected by moisture and returns to earth as acid rain.

Polluted waters, besides being unfit to drink, are harmful to wildlife. Fish and waterfowl are particularly affected by pollution. Many species of animals are endangered by it.

Noise pollution

Some of our activities create noise. Certain industrial processes are so noisy that workers must wear ear protection to protect their hearing, Fig. 12-15.

In the transportation industry, engines are noisy. Again, proper application of technology can help. Autos and trucks are required to use mufflers to reduce noise. At airports, regulations reduce the noise made by departing aircraft.

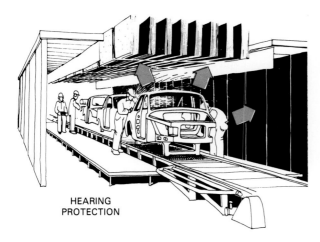

HEARING
PROTECTION

Fig. 12-15. Noise from manufacturing can be harmful. Workers in some situations wear protection to prevent damage to hearing.

Soil erosion

Farming operations and earth-moving activities can cause soil erosion (washing away or blowing away). See Figs. 12-16 and 12-17. Oil is a natural resource. It is valuable for growing food and other plants needed for their fibers. Some mining operations have also caused destruction of the land. Some mined areas have become unfit for other uses. At the same time, exposure of subsoils to the weather has caused polluting runoff.

Some soils are more fragile than others. Environmentalists and conservationists have noted

A

B

Fig. 12-16. A—Loose soil under cultivation can be washed away in a heavy rain. B—Lack of ground cover and root systems to hold soil, allows running water to gouge ditches into once fertile land.   (Soil Conservation Service)

Fig. 12-17. Winds can also erode light soils by blowing them away. (Soil Conservation Service)

Fig. 12-18. Forester inspects damage from off-road vehicles. (USDA Forest Service)

that tracks made by recreation or off-road vehicles have damaged desert lands and crop lands, Fig. 12-18. In Alaska disturbance of the soil in certain areas leaves damage that lasts for many years. Often the vegetation never recovers.

Dislocation of workers and poor working conditions

Development of new technologies has not always been good for certain people. One example is the use of assembly lines to mass produce products. Workers must work at the speed of the assembly line. They do the same tasks hundreds of times a day. This can be difficult and boring. Sometimes it has made people ill. It has caused workers to quit jobs and unions to strike for better working conditions.

Our growing knowledge has helped us to find new ways of doing things. It has resulted in new products and improved methods. Simple, repetitive tasks are being done by robots or automatic machines. But this has also caused some workers to lose their jobs or give up their homes to relocate in communities where their skills are still needed.

Using Technology Wisely

Does this mean we should not use automation? Does this mean that technology is bad? Not at all. But it does mean that we must use technology wisely. We should try to see the consequences (results) of changes before we change how we do something. Often the bad effects can be avoided. In other instances, a change can be delayed until the bad side effects can be corrected.

It is now becoming common for companies to do *technology assessment studies* to find what effect new processes or products may have on humans and all living things (plants as well as animals). See Fig. 12-19.

Fig. 12-19. Mining companies are reclaiming land disturbed by strip mining operations. This land near Gallop, New Mexico, has been seeded down. Workers blow mulch to protect the soil and provide cover for new growth. (Gulf Oil Co.)

Governments have become deeply involved in technology. Often a branch of government has set safety standards that industry must follow. For example, a government agency regulates how nuclear power stations are built in the United States. A containment shield around the nuclear core must consist of a steel shell 9 in. thick and a concrete structure over the shell 3 ft. thick. The Chernobyl reactor that blew up had no such protection.

There are many instances of regulation to preserve the environment and protect life. New automobiles must have certain pollution controls. Also, the use of leaded gasoline is being discouraged. Many states now require periodic testing of vehicles for emissions, Fig. 12-20. Certain pesticides and weed killers may no longer be manufactured because they were damaging to human and animal life. The U.S. government also attempts to protect workers through industrial safety regulations.

We can never go back to a life without our technology. Nor should we want to. Look around you. What conveniences do you have that you would want to live without? Technology is here. It will always be here. It is up to each one of us to see that it is applied wisely.

Fig. 12-20. Automakers are designing automobiles which burn fuels better. Governments are also setting up testing stations to assure that motorists keep pollution control systems in good working condition.
(American Petroleum Institute)

SUMMING UP

Technology has given us many benefits that improve our lives. These benefits improve our environment and give us more time to enjoy life. We have better food, better housing, and better health because of technology. Thanks to communication technology and transportation technology, we are able to communicate with one another and travel with ease.

However, there are some effects of technology that have not been good. We are finding that some resources are scarce. We are seeing that in using some technology we have caused pollution of air, water, and soil. Technology has also caused loss of jobs for some. It has caused other families to be uprooted as they move to other places to find jobs that were lost.

We need to use our technology more wisely to do away with or reduce the undesirable effects. Governments are attempting to regulate industrial activities that cause problems. Companies are now studying the possible effect of new technologies. Each person has a responsibility to see that our use of technology does not pollute or cause waste of nonrenewable resources.

WORDS TO KNOW

These words are used in this chapter. Do you know their meaning?

Damage to environment, Depletion, Earth berming, Erosion, Getting away, Impacts, Leisure, Nutrition, Pollution, Productive, Recycle, Scarcity, Terpenes.

TEST YOUR KNOWLEDGE
_____ Chapter 12 _____

Do not write in this text. Place your answers on a separate sheet.
1. Describe the meaning of the term, impacts.
2. List two reasons why we look for change in our environment.

3. Give one example of how technology has affected your life.
4. All of our technology is linked together; that is, one type depends on the other. True or False?
5. Before oil could be recovered from the earth, _____ had to be developed so that drills could be manufactured.
6. Being _____ means doing something that contributes something of value to life.
7. Technology has provided much work for people. True or False?
8. When any material becomes scarce we need to (indicate all correct answers):
   a. Stop using the material.
   b. Use the material more wisely.
   c. Look for a substitute that is renewable.
9. List two negative effects of the way we have used technology.
10. It is common now for companies to do _____ _____ _____ to determine what effect a new technology is likely to have on the environment.

11. It is up to the following to see that technology is wisely used:
   a. Governments.
   b. Conservation officials.
   c. Experts in various fields of technology.
   d. People like yourself.
   e. All of the above.
   f. None of the above.

SUGGESTED ACTIVITIES

1. Make a list of the ways in which energy could be saved in any of our transportation modes.
2. Watch the newspapers and magazines for articles on a way that technology is benefiting humankind.
3. Clip and bring to class articles having to do with pollution problems. Have a class discussion on the problem.

New aircraft engines are being developed for better fuel efficiency. This was the first test flight of an ultra-high bypass turbofan engine developed by General Electric Company. It is expected to revolutionize commercial aviation in the 1990s with fuel savings of 40 to 70 percent. (NASA)

APPLYING YOUR KNOWLEDGE

Introduction

The use of technology impacts our lives daily. The clothes we wear, the type of vehicle we ride in, the way we communicate, the types of product we can buy are all results of technological decisions. Also, the amount of pollution in the air and water around us is a result of technology. We must learn to control the use and abuse of technology. We must use our resources wisely. We must consider the impact of each technological decision we make.

This activity will give you a chance to consider ways to recycle waste from our technological systems.

Equipment and Supplies

Pencil and paper
Assorted scrap and waste materials

Procedure

Your teacher will divide the class into groups of 3 to 4 students. Each team will:
1. Prepare a form like the one following.
2. Survey the school to determine the type of waste that is produced in:
   a. The classrooms.
   b. The school office.
   c. The cafeteria.

3. Complete the form by:
   a. Listing each scrap or waste item located.
   b. Listing the approximate quantity of each scrap or waste item produced each week. Consider sheets of scrap paper from office copier, tin cans from cafeteria, etc.
   c. Indicating which location (classroom, cafeteria, office) produced the waste.
   d. Indicating the type of scrap or waste. The types are:
      1. Recyclable — can be used to produce like products. For example, aluminum cans can be melted to make new aluminum.
      2. Biodegradable — will rot (decompose) into a natural material when it is buried in the ground.
      3. Solid fuels — can be burned to produce heat energy.
      4. Reworkable — can be used to make other products (lumber scraps being used to make doll furniture).
      5. Solid waste — cannot be reused to make a product but can be used for landfills.
4. Select one of the reworkable materials and develop a use for it. Report to the class on your choice. This should include:
   a. A description of the material.
   b. Drawing for the product or device.
   c. A prototype of the product or device.
   d. Directions for using the device or product.

WASTE AND SCRAP MATERIAL SURVEY			
Prepared by: (1) (3)	(2) (4)		
ITEM	APPROX QUANTITY	LOCATION	TYPE
1.			
2.			
3.			
4.			

Chapter 13
Technology and the Future

The information given in this chapter will help you to:
- ■ Discuss how rapidly technology is changing the way we live.
- ■ Define the term, "futurist."
- ■ List the methods used by futurists to predict technology of the future.
- ■ Explain why we should closely examine each new technology before putting it to use.

Technology is rapidly changing our world. It is bringing us services and products beyond our grandparent's wildest dreams. It seems that with each year the pace of change quickens. Each new process or invention makes still other advances possible.

Consider the things that you take for granted in your life. You have instant communication. By pressing a button you can get news from around the world. Soon it will be possible to map distant planets and radio the information back to earth. See Fig. 13-1. An airplane can carry you hundreds of miles in an hour or more. Nuclear energy provides light and electric power in your home. Astronauts have landed on the moon.

In your great grandparent's time, a horse and surrey was a common way to travel, Fig. 13-2. Automobiles had only begun to replace horses,

Fig. 13-3. The first radio station was being formed and the invention of television was still 30 years away. Space travel was for dreamers and readers of adventure stories.

Look at Fig. 13-4. What would a pioneer moving West in 1848 think of our transportation system? And what would a Pony Express

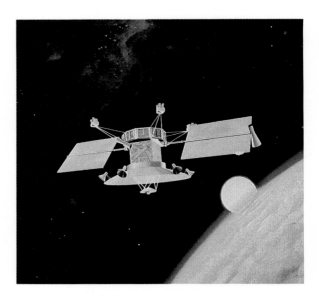

Fig. 13-1. With our "instant" communication, it is hard to imagine that television was not in general use until the 1940s. Now we are planning to map Venus and send pictures back to earth. (NASA)

Fig. 13-2. A mode of travel from many years ago was the horse and surrey. You can still find them being used in Amish communities.

Fig. 13-3. Perhaps your great grandparents were amazed when vehicles such as this truck replaced the horse and wagon. It hauled fuel and lubricants overland around 1910. (American Petroleum Institute)

Fig. 13-4. Pioneers moved west 140 years ago in covered wagons pulled by oxen. The trip took months. Now airplanes make the trip in several hours.   (Currier & Ives)

rider think of our methods of sending messages? What would an Indian cliff dweller think of our skyscrapers? How would the Wright brothers react to today's aircraft? See Fig. 13-5.

Without a doubt, today's technology would be more than they could grasp. Think then, how you might feel about the changes to come in the next 50 years! Some of those changes will be even greater than anything in the past. You will participate in the changes.

NEW GAINS IN TECHNOLOGY

We are seeing great gains in technology even as you read. One of them is in the field of **superconductivity**. (Superconductivity means that a material has little resistance to the movement of electricity through it.) Copper and silver have been the best materials for electrical conductors. Right now they offer the least resistance to movement of electricity.

But, for 75 years researchers have known that at very cold temperatures certain metals have no resistance to electricity. Even at a very low voltage, current will move easily.

Still, the information was of limited value. It is not always possible or practical to cool the metal to that temperature. It could be used only for special situations.

However, it has driven researchers to find metals that can superconduct at warmer temperatures. Little happened until December 1986. Then research found that a ceramic (sandlike) material mixed with copper and oxygen became superconductive at $-295°F$. Fig. 13-6 helps explain superconductivity.

Fig. 13-5. New twin fuselage (body) air transports may not be far from reality. Aircraft designers say they will be more efficient than today's single fuselage airplanes. (NASA)

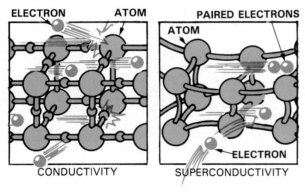

ELECTRON　　ATOM　　PAIRED ELECTRONS
ATOM
ELECTRON
CONDUCTIVITY　　SUPERCONDUCTIVITY

Fig. 13-6. How electric current would look if magnified many times. Left. Normal conductivity. The electrons (dark balls) move through a conductor with some difficulty. They bump into atoms, impurities, and other electrons. Collisions waste energy and slow the current. Right. In a superconductor, electrons pair up attracted by positive charges. Their momentum (force of motion) eliminates resistance.

For most applications it would be too hard and costly to keep conductors that cold. Imagine trying to keep wiring in a house that cold! Still, it caused a spurt of research activity. Everyone began working on the problem! In July 1987 another research team found a material that lost all resistance at near room temperature. They also discovered the crystal structure of the superconducting material. (Structure is the way molecules link together.) This information provided a kind of "recipe" for producing the material in bulk.

There is great excitement about being able to transmit electricity with no resistance at normal temperatures. It makes possible faster and smaller computers. It could give us high-speed trains. Transmission lines for electric power would be more efficient. Smaller generators could produce more electricity. They would also use less fuel or water power.

The research group also found a way to produce the superconductor on a thin film. If the two discoveries can be combined, we could have ultrahigh-speed electronic circuits. Computers would become faster and cheaper. The circuits would create little heat. Expensive cooling would not be needed.

What We're Likely to See

This is but one example among thousands. In the near future we will most likely see:
1. Under sea mining.
2. Solar-powered cars. See Fig. 13-7.
3. Interactive (two-way) national television.
4. A few "workerless" factories.

Fig. 13-7. Automakers and other companies are already experimenting with solar-powered cars. This vehicle, known as the GM Sunraycer, won the first transcontinental World Solar Challenge race in Australia.
(GM Hughes Electronics)

5. Roadways which "guide" the vehicle.
6. Electricity generated by the action of the ocean waves.
7. Hydrogen fuel extracted from seawater.
8. A "cashless" society (all payments made by computers).
9. Around-the-world airplane flights of less than eight hours.
10. Magnetic levitation trains (held above the ground by magnetic force).
11. Liquid crystal window shades. (Use electronics to make windows transparent or opaque.)
12. Electronic newspapers.
13. Computers that will give doctors three-dimensional colored "x-rays."
14. Miniature mechanical "helper" hearts that work inside the human heart.
15. LHR (low heat rejection) engines that will deliver better fuel economy.
16. Factories in space. See Fig. 13-8.
17. Auto air conditioners that work off waste engine heat.
18. Superionic conductors (conduct ions rather than electrons).

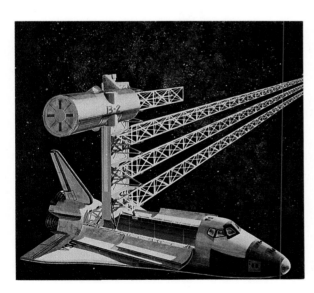

Fig. 13-8. Artist's ideas of a "space factory" making a large beam structure while in orbit around the earth. Work crews would be housed in the space shuttle oribiter.   (NASA)

How do you think people develop lists of "future" events? Are they just wild guesses? Generally they are not. They are the result of study. Let's take a look at predicting the future of technology.

Seeing into the Future

It is not too hard to predict that superconductive materials might be used sometime in the near future. But trying to imagine discoveries far into the future is harder. Clues are few. Wouldn't you like to look ahead and guess what changes are yet to come?

There are some people who specialize in study of the future. We call them **futurists.**

Futurists

Futurists are not at all like "fortune tellers." They do not have supernatural powers.

By looking at what is happening now, they try to describe what is possible. Through a study of what is changing today they make reasonable guesses about what will change in the future.

Futurists have formed organizations all over the world. There are centers in Germany, Czechoslovakia, England, Russia, Italy, the United States, Venezuela, and Brazil.

Among the futurists are economists, mathematicians, physicists, and operation researchers. They are concerned about changes that are likely to occur in years to come.

Futurists deal with questions such as:
1. When will **aquaculture** (growing plants in the oceans) be able to feed the earth's people?
2. Is it likely that electric cars will replace gasoline-powered cars in the next 20 years?
3. What changes are likely to occur in city transportation systems in the next 30 years?

There is practical value in what futurists do. They give us clues about changes that will come. Clues help us to choose a course of action. For example, when the first gasoline engine was developed, a futurist might have predicted that, in 25 years, engines would pull buses. This information would have told people what kind of new jobs were developing. It would have started buggy factory owners building engines and automobile bodies.

How do you go about studying the future? The future hasn't happened yet!

Futuring Methods

Futurists use special methods. Most of them involve making **projections**. Projections are educated guesses about what may occur. Some projections are long range. Others are not.

A short-range projection deals with what can happen in the next couple of years. Looking 10 years into the future would be a mid-range projection. Long-range projections look ahead as far as 50 years.

Methods of predicting the future include:
1. Collecting information about trends.
2. Use of analogies.
3. Taking surveys.
4. Networking and decisions.
5. Developing a "future history."
6. Scenarios.
7. Modeling.

Trends

Studying **trends** means looking at a certain activity. The study trys to see what general direction events are moving. For example, in the last few years automobiles have become smaller. Engines are less powerful. Passenger space is not as large. We can call it a trend. Futurists use other facts to predict how long this trend will last. Trends in sizes of families will be considered. So will the price of gasoline or fuel scarcity. See Fig. 13-9.

Suppose that your class was looking for a product to make for sale to other students. Your study of student interest might show more and more students were becoming interested in skateboarding. This could be a trend. It will influence the class decision. You might decide to design and build skateboards.

Analogies

An **analogy** is like a comparison. We see things or events that are alike in some ways. A futurist assumes that what happens in one situation, may also happen in a similar situation, Fig. 13-10.

This is how an analogy works:
1. "Our supplied of petroleum are scarcer and more expensive. Indeed, we already have to pay high prices to foreign countries for petroleum. We need an inexpensive fuel to replace petroleum."
2. "But something similar happened around 1942. There was a shortage of rubber during World War II. It led American technology to develop a synthetic rubber."
3. "It is reasonable that American technology will develop an affordable replacement for petroleum."

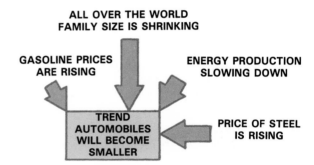

Fig. 13-9. Futurists will study events that are happening. They can predict the effect of these events.

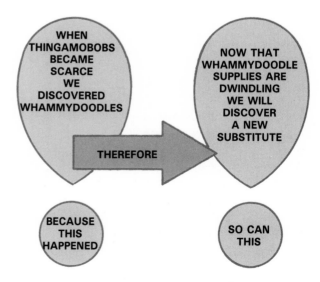

Fig. 13-10. This is an example of an analogy. If one event occurs, so can a similar one.

Surveying

In a **survey** futurists ask a number of experts a series of questions. Then they compare the answers. For example, they may want to find out how computers might be used in automobiles five years from now. They could ask a number of automotive engineers for their opinions.

Delphi study

A **Delphi study** is a special kind of survey. A group of experts respond to a question about the future. Their answers are only their best guesses. But the survey does not end with the first response. The guesses are written down and shared with the experts. They are asked to guess again. The questioning continues until the experts reach agreement on a prediction.

Networking and decisions

Networking and decisions is a method of making things happen. This is how it might work for you:
1. Your family decides to cut electricity and heating costs in your home by 10 percent.
2. List all the ways you can save energy.
3. Write the cost of each item you have listed.
4. Then you decide which of the cost-cutting methods you want to use.
5. You will place cost-cutting measures in the order you wish to do them. You might do this on a chart.
6. At the last step in the chart you will reach your goal.

Future histories

In the **future history** method you image yourself being alive far into the future. Then you look "backward" and describe events that lead to the future time you have picked. Science fiction writers use this method. Fig. 13-11 is an example of a future history. This approach gives the imagination great freedom. It is free of the predictions of "experts."

Scenarios

In using the **scenario** method, futurists imagine what could happen. First, the scenario describes a set of related events. Then it gives the consequences of those events. Usually more than one scenario is developed for the events. Fig. 13-12 is a scenario for a different city from the ones we know. Can you think of other consequences from these same events?

Modeling

Modeling means making up a chart of the key parts or elements of a major event, Fig. 13-13. The charting allows the planners to predict all parts of the major event.

Suppose that a club to which you belong is planning a dance. The dance is the major event in the model. Other minor events are necessary before the major event can occur.

In constructing a model, the club members chart all the minor events:
1. Hiring a band.

Here is it, 2099. As is my custom I sit here at my mindscriber memory writing my thoughts on the great progress we have made in technology. Early in the century—about 2010—we made amazing discoveries in how the human mind works. From this work came advances in cybernetics. (This is how machines and people can work together.) In the 1970s and 1980s computers took over many arithmetic functions and lesser management functions of humans. This led to the new technology of cybionics. New machines are closely linked to our thought and speech patterns. First, the "speakwriter" took over as a word processor. Sensitive transducers could translate speech into printed books and magazines. Later models could "paint" pictures as they were described by the speaker. This revolutionized the communications industry. Still another breakthrough came in 2040. Machines like my mindscriber tapped into human brain waves. By analyzing the minute electric impulses, the machines translate thoughts into voice communication or printed words.

Through microwave transmission, reporters simply view news as it is happening and broadcast simultaneously throughout the world without speaking a word.

Technologists and scientists are working on a new transporter which they have named the "digestomobile." Since world governments have banned use of fossil fuels it will replace gasoline-fueled transporters. The heart of the vehicle is the digester which takes in garbage and converts it directly to energy to drive an electric motor. When it is perfected, landfill problems that have plagued us since the 1980s will disappear.

Fig. 13-11. This is an example of a future history. It is highly imaginative. Yet, it is based on some factual information.

By the year 2040, city planners found that our major cities needed to solve serious problems. Moving about in the central city was a problem. Streets were clogged with traffic. Expressways had not helped the problem to any degree. Even worse, air pollution from the traffic was threatening the health of citizens.

Planners decided on a bold plan to reduce traffic. First, all personal vehicle traffic was banned from the cities. Mass transit systems replaced the automobile. Only trucks delivering goods, foodstuffs, and services were allowed within city limits. Bicycle traffic was allowed anywhere.

Neighborhood shopping areas were located where they were within easy walking distance for shoppers. Streets were no longer congested with traffic. Air pollution was reduced. Many commuters were attracted back to the central city by the quality of life they found there.

Fig. 13-12. A scenario is also an imagined event. It looks at related happenings and attempts to predict what the solution might be. Usually, several scenarios will be written for the same set of events.

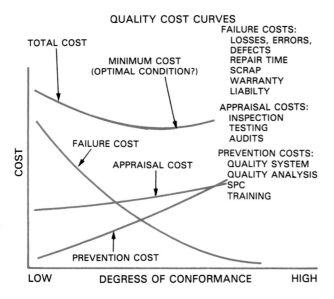

QUALITY COST CURVES

Fig. 13-13. This is a model of a company's costs for a quality control program. The major parts (events) of the program are listed at the right. These events are plotted in the lines at the left. They show that as more is spent on prevention costs, total costs come down. But at one point, spending more on prevention does not lower total cost any more. Why do they call the lowest point of total cost the "optimal (most desirable) conditions"?

2. Renting a dance hall.
3. Hold a fundraiser (carwash, candy sales, etc.) to pay for the band, hall, and decorations.
4. Decorate the hall.
5. Plan publicity for fundraising.

If any of the events are changed, planners can reschedule the other events. The hall could be donated. The band could donate their time. This would allow the fundraising event to be dropped from the model.

CONCERNS ABOUT TECHNOLOGY

The effects of technology have not always been good. You learned this in Chapter 12. Often new methods were introduced with little or no idea what they might do to people or the environment. The Romans are a good example. They piped water into their homes using lead pipe. The lead slowly poisoned the people. Many important families got sick and died. Some experts say this was one reason for the decline of the Roman Empire.

In our own society, we began to use large quantities of pesticides and herbicides. They controlled insects and weeds. Weed killers were a benefit because they increased crop yields. The pesticides helped prevent insect-borne diseases. We had cheap and plentiful food. But later, we discovered that the side effect of pesticides and herbicides were harmful to all forms of life.

Another example: fishermen in the North Sea recently improved fishing technology. It greatly increased catches of herring. However, no thought was given to the future of herring fishing. Now the herring are disappearing. As the herring disappear, so are fishing fleets. So are whole fishing villages.

The technology was good, but humans did not know how to use it well. We are beginning to learn that we must live in harmony with nature. We must realize that resources are precious. They must be used wisely.

In the future, side effects of any new technology will have to be studied. We are becoming more cautious. We cannot accept new technology without question.

There are many indications that we are learning to make better use of our technology. We are using it to economize (save). For example, until recently, telephone communications used up miles of thick copper wire. Fiber optics will do away with this. The hair-thin fibers are made of glass or plastic. They will carry the messages in the future, Fig. 13-14. Compared to copper, the fibers require only one-thousandths the energy to produce. The same energy that produces 90 miles of copper wire would turn out 80,000 miles of fiber.

We have become cautious. The Congress of the United States has set up an Office of Technology Assessment. Staff members study the effect new discoveries are likely to have on people and the environment.

We have become most concerned about acid rain, ozone depletion and global warming of the climate. We may need to make a hard choice between fossil fuels and nuclear fission for producing electric power. While nuclear power presents the danger of radiation, it does not contribute to acid rain, ozone depletion, or the greenhouse effect (warming of the earth because of putting carbon dioxide into the atmosphere).

SUMMING UP

Technology is rapidly changing our lifestyle. We take modern travel and communication methods for granted. They are commonplace. Those who lived during the 1890s and even into the 1920s would be amazed at the changes that have occurred since.

Change will continue to take place at a rapid pace. It would be interesting to predict what might happen in the next 50 to 70 years. Such predictions are becoming a science. Those who work in this field are called futurists. The futurists use seven different methods in their forecasting. All are based on observing what is happening in the present.

In the future we will need to examine our new technology carefully to assess its possible ill effects. This will help society avoid mistakes of the past.

WORDS TO KNOW

These words are used in this chapter. Do you know their meaning?

Analogies, Decisions, Delphi study, Fiber optics, Future history, Futuring methods, Futurists, Modeling, Networking, Projections, Scenario, Superconductivity, Surveying, Trends.

SUGGESTED ACTIVITIES

1. Imagine yourself living 100 years from now. Describe what it is like to be in school. Write a "future history" about a day in the school of the future.
2. Write a paper on how the gasoline engine led to the invention of the first automobile. Ask your instructur to suggest where you might find information.

Fig. 13-14. Fiber optics are making telephone installations cheaper and better. Light-carrying glass or plastic fibers transmit the voice. (AT&T Bell Laboratories)

3. Imagine what a futurist would have done with the information that a German inventor had built a working model of a gasoline engine.

TEST YOUR KNOWLEDGE
_____ Chapter 13 _____

Do not write in this text. Place answers to test questions on a separate sheet.
1. Give one reason why superconductivity was important news in 1987.
2. A _____ is someone who predicts certain technological developments.
3. Predicting the future discoveries and technologies:
   a. Is done for the fun of guessing what will happen.
   b. Is very inaccurate.
   c. Is worth the attempt because it helps us to plan for change.
   d. Is useful only for those in business or manufacturing.

4. List seven ways of studying what inventions or discoveries will be made in the future.
5. It is wrong to question the use of new technology. True or False?

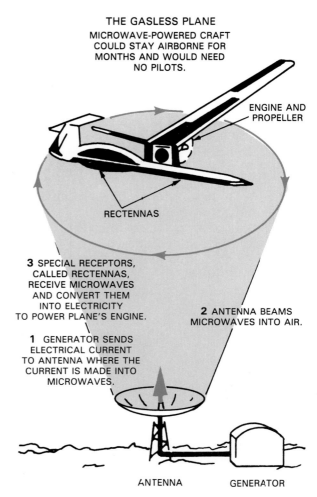

THE GASLESS PLANE
MICROWAVE-POWERED CRAFT COULD STAY AIRBORNE FOR MONTHS AND WOULD NEED NO PILOTS.

ENGINE AND PROPELLER

RECTENNAS

3 SPECIAL RECEPTORS, CALLED RECTENNAS, RECEIVE MICROWAVES AND CONVERT THEM INTO ELECTRICITY TO POWER PLANE'S ENGINE.

2 ANTENNA BEAMS MICROWAVES INTO AIR.

1 GENERATOR SENDS ELECTRICAL CURRENT TO ANTENNA WHERE THE CURRENT IS MADE INTO MICROWAVES.

ANTENNA   GENERATOR

Your great grandfather may have ridden in one of the first electric cars. Modern technology was able to convert it to solar electric power. What will happen to auto transport in your lifetime?

In September, 1987, researchers from the Canadian Communication Research Center flew a small gasless plane in Ottawa, Canada. It was the first step in a communication satellite program to use microwaves to keep the satellite aloft. A generator sends an electrical current to an antenna which turns the electricity into microwaves. Antenna beams microwaves to the circling plane. Receptors in the plane collect microwaves reconverting them to electricity. The electric current drives the electric motor connected to the propeller.

APPLYING YOUR KNOWLEDGE

Introduction

Technology has given us the life we have today. This life is very different from what our grandparents had. Likewise, life for future generations will be very different from life today. People may be living in deep space or in undersea settlements.

This activity will allow you to consider a life in space. You will design a self-contained space station. Look at Fig. 13A.

Equipment and Supplies

No. 10 tin cans (large can found in most school cafeterias)
Paper towel tube

5 — 6" cardboard or posterboard squares
1 1/2" wide strips of posterboard
Compass
Ruler and pencil
Scissors or small tin snips
Glue
Masking or transparent tape
Wall paper samples, scraps of cloth, etc.

Procedure

Your teacher will divide the class into groups of 3 to 4 students. Each group should:
1. Obtain the materials listed above.
2. Study the Figs. 13A and 13B to determine the basic structure of the "space station."

Fig. 13A. Drawing of what might be a typical space station. The module has been split in half. Left half has a command and control center, an airlock for access to space, a galley (kitchen), sleeping quarters, and an observation deck. Right half contains the other half of the observation deck, more sleeping quarters, exercise and recreation areas, storage for space suits, and the other half of the command and control center.Spaces above and below living quarters are for mechanical systems.   (NASA)

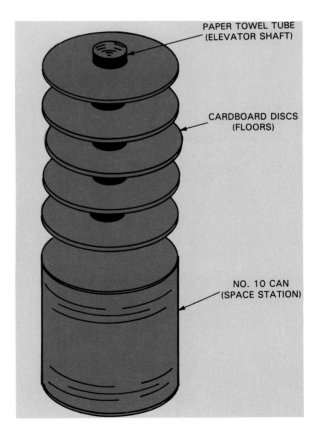

Fig. 13B. Basic structure of the model space station.

3. List the basic areas needed for complete, comfortable living. These may include:
   a. Living quarters.
   b. Recreation areas.
   c. Agriculture, production, or factory area (the reason for the space station).
   d. Mechanical support areas (heating, air conditioning, waste disposal, etc.).
   e. Office or managerial areas.
   f. Flight control areas.
   NOTE: Be sure to consider other types of facilities. This is not a complete list!
4. Assign each basic area to one of the four floors of the space station.

Each member of the group should select one or two floors to develop. He or she should:
1. Cut a 5 7/8 in. disc from a square of cardboard or posterboard.
2. Cut a 1 1/2 in. hole in the center of the disc.
3. Draw a floor plan on the disc. Be sure to consider the wise use of space, ease of movement from different areas, and grouping like activities together. Remember, each inch equals eight feet.
4. Cut walls from the 1 1/2 in. strips of posterboard.
5. Cut doorways in the wall sections.
6. Attach the walls to the floor.
7. Decorate the walls and floors with wallpaper samples and cloth to represent actual surface treatments.

The group should complete their space station by:
1. Cutting openings in the elevator shaft (paper tube).
2. Attaching the floors to the elevator shaft. See Fig. 13B for the spacing.
3. Insert the space station into the launch shell (tin can).

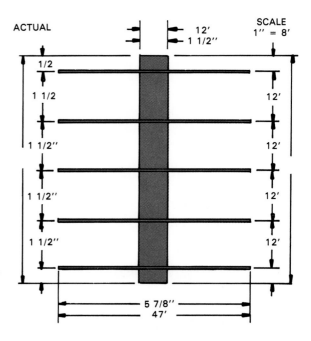

Fig. 13C. Elevation view of the space station.

Technical Terms

A

Acoustical properties: qualities of a material which govern how it reacts to sound waves. Materials that absorb the waves are said to be insulators. Materials that carry sound are called transmitters.

Actuators: devices in modern engine control systems which use a small electric current to cause movement. They can be small electric motors or solenoids (magnetic switches). They can open or close electric circuits and control valves.

Age of information: a period of time when the main industry of a country is the collecting, processing, and exchanging of information.

Alternating current: current which changes direction of flow at regular intervals called cycles.

Analogy: a comparison of things in certain circumstances which have a likeness in one respect. They may be quite different in other respects.

Aquaculture: using the ocean's waters for growing of crops.

Armored cable: electrical conductors encased in a flexible metal covering. Also called "BX."

Assembly drawing: drawings made to show how to assemble a product.

Audience assessment: gathering information about a group of people to be reached by a message.

Audio recordings: the recording of sound, such as talking, singing, or music, on tapes or records. (Audio means sound.)

B

Bill of materials: a list of all the parts needed to make one product. It gives the part name, size, and quantity of material to be used.

Biomass: the sum of all organic matter in an area.

Boundaries: property lines established by a survey.

Brainstorming: a group process for solving a problem. The group is asked to think of as many solutions as possible in a short time. They are also encouraged to expand on one another's ideas. They are discouraged from criticizing or evaluating any of the ideas at this stage.

Broadcast messages: communication carried by radio and television stations.

Building permit: document issued by a city or other community allowing work to proceed on construction. It is issued after submitting an application together with a set of building plans and specifications that meet local building codes.

C

Career ladder: the process one goes through to work his or her way up to better and better jobs through experience, additional training, or education, and perseverance.

Cargo: material or products carried on transportation vehicles.

Carrier: a channel for carrying electronically

generated messages. It can be radio waves of the electromagnetic spectrum, a wire conductor, or optical fibers.

Ceramic: inorganic (never living) matter made up of cyrstals. Types include: clay-based, refractories, and glass.

Chain: a surveying instrument consisting of a steel measuring tape. It measures distance along a boundary line.

Chemical energy: reaction between two substances when mixed. For example, when petroleum and oxygen are mixed they will burn rapidly if ignited.

Chemical processing: using chemicals to change the form of materials. The chemicals change the structure of the material's molecules.

Chemical properties: that which controls how a material will react to chemicals. For example, some materials will corrode or form rust. Others will resist corrosion.

Civil structures: structures designed for public use including, roads, bridges, monuments and other public projects such as sewers, pipelines, etc.

Collar beams: boards which brace rafters, preventing them from pushing walls outward.

Commercial, public, and industrial buildings: structures that give people shelter for certain activities such as shopping, working, traveling, etc.

Communication: the sending and receiving of messages.

Communication technology: use of equipment and systems to send and receive information.

Composite: a solid material which combines two or more materials. Yet, each material retains its own properties. Concrete is an example. It combines cement, sand, and gravel but none of the ingredients change.

Computer model: three-dimensional view generated on a computer screen. Some computers can actually "test" the product before it is built.

Condemnation: a process which allows governments to acquire real estate when the owner is not willing to sell.

Conduit: lightweight pipe designed to enclose electrical wiring.

Conservation: making better use of the available supplies of any material.

Contractor: one who hires workers and directs building processes.

Conveyors: stationary structures that move material. The structure supports rollers or belts over which the material is transported from one place to another.

D

Damage to environment: polluting or depleting natural surroundings so as to reduce the quality of life.

Database: all that is known about any subject. In computer systems, stored information that has structure and organization. Also, it can be called up and used.

Date processing: process of entering data (facts) into a computer. Also includes giving the computer instructions to follow and receiving the results of the computer's operation.

Decoding: putting a meaning to a message; that is, understand the message so that proper action can be taken.

Delphi study: a special kind of survey directed at experts in the field of the study. It involves repeated surveying and then feeding back the results to get a new opinion. This continues until the group comes to agreement on a single answer.

Depletion: the using up of a resource (petroleum, for example).

Design: plan for a device or product. The process of design begins with defining a problem. It also involves gathering information, developing and testing, finding solutions, retesting, and selecting the best solution.

Detail drawing: generally, orthographic (two-dimensional) drawings which give the size and shape of an individual part. Manufacturing workers will use this drawing to make the part.

Detailed sketch: sketch which contains detailed information on size and shape of the product. It will show length, width, squareness, roundness, etc.

Directors: persons who sit on the governing boards of companies. They set major policy, appoint company officers, and set their salaries.

Distance multiplier: a simple machine, such as a lever, which can change the amount of movement applied to it. A small motion applied at the end of its arm will create a greater movement at the end applied to the load.

Dot matrix printer: a printer that forms letters and symbols by placing tiny dots on paper to form the letters and symbols.

Drywall: a wall covering consisting of a chalky material contained between two layers of heavy paper. It is applied in 4 by 8 ft. sheets using nails or glue.

Ductility: ability of a material to be pulled, stretched, or hammered without breaking.

E

Earth berming: banking (piling) soil around one or more sides of a building to take advantage of the insulating qualities of the ground.

Earth sciences: branch of science concerned with history, properties, composition, and behavior of the natural world. Includes meteorology, oceanology, and geology.

Eave: low end of a pitched roof.

Electrical code: rules governing installation of electrical wiring. For reasons of safety, it specifies both materials and methods of electrical wiring.

Electrical energy: the energy of moving electrons. The movement can be caused by lightning, batteries, and generators. Moving electrons create an electric current.

Electrical generator: device which can convert mechanical energy into electrical energy. It is an energy converter.

Electrical/magnetic properties: that which governs how a material will react to electrical current. For example, some materials will carry current while others will not.

Electrical properties: qualities which control a material's reaction to electrical current.

Electromagnetic induction: a physical principle which causes a flow of electrons (current) when a wire moves through a magnetic field.

Electrolyte: a chemical solution through which an electric current can travel.

Electromagnetic radiation: energy that moves through space in waves. Heat and light from the sun are two examples.

Encoding: all the tasks that are part of designing a message to be sent through a communication system. The message may be a symbol, sound, or motion.

Engineering material: solid matter which has a set, rigid structure. Solids maintain this structure without support from a container.

Entrepreneurship: the starting and managing of a business for the purpose of making a profit.

Estimators: persons who estimate the costs of time, labor, and materials in a construction project.

Excavating: digging holes and trenches in the ground for some part of a structure.

Exhaustible resources: material which, once used, can never be replaced. Examples are fossil fuels such as coal and petroleum.

External combustion engine: a heat engine which burns fuel outside the engine.

F

Feedback: data on how well a system is operating. It is designed to provide information so adjustments can be made to improve the outputs of the system. Also: In the design process, use of information from a later step to improve an earlier step. Purpose of feedback is to improve the operation of any system.

Film messages: photographs and transparencies, movies, slides, and filmstrips used for entertainment or for presenting information.

Flexography: type of printing done from a rubber-like plate with raised letters on it.

Force multiplier: quality of a simple machine which makes it capable of exerting more force to a load than is applied to the simple machine.

Format: size and shape of a printed message.

Forms: temporary structures built to contain concrete until it hardens.

Foundation: that part of a structure which ties it to the ground. It also supports the weight of the structure.

Four-cycle engine: heat engine having a power stroke every fourth revolution of the crankshaft.

Freighters: ocean-going vessels that are designed to carry products and materials.

Fuel cell: an energy converter which converts chemicals directly into electrical energy.

Future history: a method of determining what might happen in the future by writing down an imaginary story about it.

G

Geothermal energy: energy originating deep inside

the earth in the form of heat.

Girder: a large beam, usually horizontal, made of wood or steel. It supports the joists (floor frame) of a building or the superstructure of a bridge.

Goals: reason or purpose for a system.

Graphic communication: messages such as pictures, graphs, photographs, or words which are placed on a flat surface.

H

Hardware: equipment/components that make up an entire computer system.

Head: In a hydroelectric power station, the distance from the water level down to the turbine.

Heat energy: a form of energy. It is present in the increased activity of molecules in a heated substance.

Heat engine: an energy converter which converts energy such as gasoline into heat and from heat into mechanical energy.

Helicopters: aircraft designed to land and take off in very limited space. Lift is supplied by a rotor which takes the place of wings. A second rotor is used for control.

I

Inclined plane: a surface placed at an angle to a horizontal surface. It makes possible the moving of a heavy object from one level to another with less force than lifting straight up.

Input device: any piece of hardware, such as a keyboard, that can be used to enter information into a computer.

Inputs: resources used by a system.

Intermodal transportation: systems of travel which use more than one transportation system. For example, an air traveler may use land transportation to an airport, and an escalator or people mover (moving sidewalk) to get to the plane.

Internal combustion engine: a heat engine which burns fuel inside its cylinders.

L

Leisure: time free from work during which people may rest or take recreation.

Lever: a mechanism or simple machine that multiplies the force applied to it. It consists of a long arm to which the force can be applied and a fulcrum (pivot point) on which the arm rotates.

Life sciences: study of living matter. Branches include: botany, zoology, microbiology, medical science, and agricultural science.

M

Machine: simple machines are devices or mechanisms which change the amount, speed, or direction of a force. Also: a framework to which mechanisms or tools can be attached to make work more efficient.

Magma: heated molten matter trapped deep beneath the surface of the earth.

Materials: substances from which useful products or items are made. Technological systems use three major types of materials:
1. Substances which provide energy such as petroleum, coal, and wind or falling water.
2. Liquids, gases, and nonrigid solids that support life. Air, water, and fertilizers are of this type.
3. Industrial engineering materials. These are solids with a rigid structure. They are the basis for all products having a set form.

Material property: a characteristic which a material has. It affects how a material reacts to outside conditions. There are seven groups of properties: physical, mechanical, chemical, thermal, electrical and magnetic, optical, and acoustical.

Mechanical processing: changing material by cutting, crushing, pounding, or grinding into a new form.

Mechanical energy: the energy present in moving bodies. It is sometimes called kinetic energy.

Mechanical properties: qualities of a material that affect how it reacts to mechanical force and loads. They affect how it will react to twisting, pulling, and squeezing forces.

Mechanical systems: provide convenience and comfort inside structures that provide shelter.

Mechanism: a basic device that will control or add power to a tool. They are designed to multiply the force applied or the distance traveled. The six mechanisms are: the level, the wheel and axle, the pulley, the inclined plane, the wedge, and the screw.

Media: any material that will carry a message from a sender to a receiver. Examples are paper, film, and airwaves.

Metal: inorganic (never living) material which is usually in a solid form. Other marks of a metal are its opacity (can't see through), ductility (easily stretched or shaped), and conductivity (easily conducts heat and electricity).

MHD generator: a device for converting heat energy directly into electrical energy.

Middle managers: in a company or corporation, persons who have management level positions below the officers of the company. Supervisors and heads of departments.

Mock-up: an appearance model of a product. It shows how the object will look in real life but it does not operate. It is made of easily worked materials like clay, wood, or cardboard.

Model: a three-dimensional (has width, length and height) copy of a new product.

Modeling: predicting future events by charting what is likely to happen. The person decides what the major event is likely to be. Then he or she sets down or "charts" the minor events leading up to the major event.

N

Negotiate: to bargain and thus try to reach agreement. In respect to construction, the process of reaching an agreement over price and other aspects of real estate.

Networking and decisions: a method of reaching a goal. It involves group decisions on what the goal is to be and then agreeing on steps needed to reach the goal.

Nonmetallic cable: plastic-covered insulated cable containing two or more electrical conductors.

Nuclear energy: heat released when atoms are split.

Nuclear fission: a splitting of atoms of a certain kind of uranium. It releases huge amounts of heat energy.

Nuclear fusion: a reaction in which atoms are thrown together so violently that they remain joined or fused. This occurrence releases huge amounts of heat.

Nutrition: proper kinds of foods in quantities that promote growth and replacement of worn tissue.

O

Ocean liners: large ships designed to carry passengers on sea voyages.

Optical properties: that which governs a material's reaction to light. Some materials absorb light; some reflect light; others allow the light to pass through.

Output device: in computer science, a piece of hardware which allows the user to get information out of a computer. For example, a printer is an output device which can produce a typed message from the computer.

Outputs: the results of any system. Examples: manufactured product, like clothing; constructed work, like a house; communicated message, like a newspaper or a book; or a transported person, like an air traveler.

Oxidation: reaction of matter with oxygen.

P

Photographic communication: use of photographs to carry a message.

Photographic systems: a method of storing visual material by capturing it on film and paper which has been made sensitive to light. A camera captures the picture and exposes the film. After developing, the film is used in an enlarger to project the picture onto light-sensitive paper.

Physical characteristics: qualities related to one's body, such as strength, endurance, etc.

Physical sciences: science that deals with matter in its purest form. Includes physics, astronomy, chemistry, and, sometimes, mathematics.

Physical properties: basic features of material such as density, moisture content, and smoothness.

Pipeline: large-diameter pipe usually laid underground. It is designed to move liquids and loose, solid material from one place to another.

Plate bearing test: a building site test where measurements are taken to see how far weights sink into the ground.

Plates: horizontal parts of a wood wall. They secure the vertical parts known as studs.

Polymer: organic (once living) matter usually made from natural gas or petroleum. However, wood can also be changed into a polymer.

Potential energy: energy at rest but able to do work.

President: in a business enterprise, the highest-ranking member of the management team. Also: the chief company officer.

Printing systems: methods and machines used to produce words and pictures on untreated paper and other materials.

Processes: the actions taken to put resources to use. Steps taken to produce products and services.

Processing: changing of material to make it more useful.

Productive: producing something useful for the betterment and enjoyment of life.

Projections: guesses or estimates about what may occur in any situation in the future.

Project manager: person who assists contractor in coordinating the construction project.

Property: a characteristic of a material. There are seven properties: physical, mechanical, chemical, thermal, electrical and magnetic, optical, and acoustical.

Prototype: working model of a product. It is made to test the design. Usually, a prototype is made from the material that will be used in the manufactured product.

Published message: messages that are produced by printing. Published messages are produced in newspapers, magazines, books, greeting cards, etc.

Pulley: small wheel with a grooved rim. A rope is run through the wheel to lift weights. Sometimes several of them are used together to multiply applied force. Also: a wheel (attached to a shaft) that transfers motion by way of a belt.

R

Radiant energy: the energy (movement) of atoms present in sunlight, fire, and any matter.

Rafters: sloping members of a roof frame which run between the ridge board and the outer walls of a building.

Receiver: an electrical or electronic device which gathers and processes messages generated by a transmitter.

Recycle: to collect used products and reuse the materials in them. Also: finding secondary uses for manufactured produces without remanufacture.

Refined sketches: freehand drawings which add detail and develop design ideas shown in rough sketches. Their purpose is to narrow down and improve the promising solutions presented in the rough sketches.

Refining: a process that separates petroleum into various liquids or gases. These can be used for heating buildings, powering vehicles, and providing process heat for manufacturing.

Rendering: a colored or shaded sketch. It shows the final appearance of a product.

Renewable energy: an energy which will always be available because it can be replaced.

Renewable resource: material that can be replaced. Examples are growing things such as plants, animals, and sunshine.

Residential buildings: structures used for living.

Ridge: highest point of a pitched roof.

Ridge board: part of the roof frame located at the ridge. It supports and spaces the rafters.

Rockets: vehicle designed to travel in outer space where there is no oxygen. Consequently, the vehicle must carry oxygen as well as fuel to support combustion and human life.

Rolling stock: all types of railroad vehicles including engines, cars, and maintenance vehicles.

Rough in: placing pipes, ducts, and electrical wiring inside walls prior to installing wall coverings.

Rough sketches: a pictorial (picturelike) freehand drawing that shows only basic ideas of the size and shape of a product. They are done quickly and without detail to capture ideas that come to the designer.

S

Satellite: a human-made object put in orbit around the earth, moon, or another heavenly body. A communication satellite contains electronic equipment for receiving and transmitting communication signals.

Scarcity: state of being in short supply. A condition in which there is not enough of an item to meet the demand.

Scenario: a method of "futuring" in which the person sets up a series of related events that might happen in the future. Then the consequences (results) of these events are given.

Science: study of the natural laws which govern the universe.

Scientific method: a logical way of discovering new scientific information or data, or testing old scientific theories.

Screw: a simple machine in which an inclined plane is wrapped around a shaft. They are force multipliers used to fasten parts.

Sea lanes: also called shipping lanes. The water routes habitually traveled by ocean-going ships.

Sensors: devices which can sense a condition such as in an engine. When the condition is present, they send an electrical, optical, or pneumatic signal to a computer.

Sill plate: wood pieces attached to the top of a foundation. They provide a means of fastening the superstructure to the foundation.

Software: a set of instructions used to run a computer. Also known as "programs."

Specifications: written directions for the builder which control the quality of work and materials on a construction project.

Storyboard: the "layout" for a television commercial. It consists of sketchs of each shot accompanied by the script for the commercial.

Studs: vertical members of a wall frame. Usually made of wood 2 in. by 4 in. thick.

Superconductivity: a characteristic of certain materials which causes them to offer little resistance to the movement of electricity.

Superstructure: that part of a structure placed above and upon the foundation.

Survey: the act of collecting facts about a subject by questioning a select group of people.

Surveying: a method of measuring to determine boundaries of lots and acreage.

System: a set of related parts. Together they form a whole that is designed to accomplish some purpose.

T

Tankers: large ocean-going vessels that carry liquid materials such as petroleum and chemicals.

Technical means: resources such as tools, knowledge of processes, that are directed toward producing products and services.

Technician: a specialist who works with an engineer or a scientist. In engineering occupations they work between the engineer and the skilled artisan.

Technological system: an organized grouping of inputs, processes, outputs, feedback, and goals. Purpose of the system is to produce goods and services.

Technologist: a specialist in a manufacturing enterprise or some other enterprise. He or she works under an engineer or a scientist.

Technology: use of knowledge, tools, and systems to make life easier and better.

Telecommunication: system of transmitting messages over a long distance using electricity and electronics.

Telegraph: a system for transmitting messages by electrical impulses over wires or radio waves. It uses a code consisting of short and long signals called dots and dashes.

Test pit: hole bored for purpose of determining how well soil on a building site will support the intended structure.

Thermal properties: characteristics of a material that cause it to react to being heated. Heat will expand some materials. It will travel through some materials but not through others.

Thermal processing: using heat to change material into a more useful form.

Thermal properties: a material's reaction to heat. Heat may expand some materials. Certain materials allow heat to travel through them easily. Others resist movement of heat.

Tools: manufactured devices designed to perform specific tasks. Generally, it means an instrument that is worked by hand.

Transducer: device which converts heat into an electrical current or signal.

Transformer: device which can increase or decrease the voltage (force) of electric current.

Transit: a surveying instrument that measures the angle a boundary line makes with true north.

Transmitter: an electrical or electronic device that develops and sends messages over air waves.

Transmitting: any method of moving a message or signal between a sender and a receiver.

Transportation: movement of people and material from one place to another.

Trends: the general course of events. In technology, the direction taken by inventions and the producing of new products and machines.

Tugboat: small vessel which is to ships what locomotives are to trains. They push or pull loads in the water.

U

Unit train: a train which carries only one product to a single destination. For example, a train may carry only coal from a mine to an electrical power station.

Utilities: services such as electricity, water, and fuel that are brought into a building.

V

Vice-president: a company officer next in rank below a president. Usually, such a person, in a larger company, is in charge of a department of the company.

Video recordings: capturing motion pictures to be replayed through a television set. One method is to place the material on magnetic tape.

W

Waterhead: see HEAD.

Wave communication: electronic communication which uses radio waves of the electromagnetic spectrum to carry signals. Sounds or pictures are changed into pulses of electrical energy. These pulses are carried through the air by the electromagnetic waves. They are picked up at a distant point and changed back to sound and/or pictures.

Wedge: a simple machine that combines two inclined planes. It is the basis for a number of handtools including the knife, chisel, and ax.

Wheel and axle: one of the six simple machines. It is a type of lever. The axle or shaft is the fulcrum (pivot point) and the wheel is the lever.

Index